Science for Seekers

A Rationalist's Guide to God, Mysticism, Cosmic Consciousness, Free Will, Life After Death, and the Spiritual Power of the Real World

Andrew Brodsky

ISBN # 9798714816888

Cover and interior design by Andrew Brodsky

Amazon KDP Edition No. 2

www.AndrewBrodsky.net

To Lisa, who always believes in my wildest dreams.

Contents

Preface

We live in revolutionary times.

Some of these revolutions are toxic, like the war against our shared values of reason, democracy, and truth itself. There are other, more hopeful revolutions: against bigotry; against environmental catastrophe; against narrow thinking and dogma; against hate.

I believe we are in need of a new type of revolution: a spiritual one. Rationalists like me have lost religion, and we need to find it again, but we need to do so on our own terms. We are thirsty for a faith that does not require a choice between dogmatic religion, fuzzy pseudoscience, or cold reductionism. We need a belief system that respects science and admits magic; that finds God nestled snugly within the reality of the natural world; and that places the miracle of awareness right at its center. I call this way of thinking *mystical naturalism*, and *Science for Seekers* is a user's guide to the revolution I envision.

As I write this, in early 2021, my country is emerging from the shadow of two plagues, one biological and one political. Perhaps no year in recent history has more clearly demonstrated the importance of reason and science in realizing the best of humanity, and the perils of abandoning them. My interest in these pages, though, is less political than personal. It is our own spiritual growth that provides the foundation for shared values that can change the world.

Science for Seekers is the result of my own journey for meaning. I hope it inspires you to embark on a seeker's quest of your own, wherever it may lead.

Andrew Brodsky
Longmont, CO

March 2021

Science
for
Seekers

CHAPTER 1

An Introduction to Scientific Seeking

Pardon all runners,
All speechless, alien winds,
All mad waters.

Pardon their impulses
Their wild attitudes
Their young flights – their reticence!

When a message has no clothes on
How can it be spoken.

- Thomas Merton

F rom the top of Bear Peak I can see miles in every direction. To the west, the glaciers of the Indian Peaks glint in the morning sunlight. To the north and south, the ridge that marks the boundary of the Rocky Mountains extends into the hazy distance. And to the east, the entire city is arranged below me in a green grid, with the expanse of the Great Plains beyond.

On a clear June morning like today, no place in the world is as beautiful as Boulder. From my perch on a sharp pile of boulders that comprises the top of the mountain, I survey the commanding view of my adopted hometown. Among the usual collection of houses and schools and big box stores are the institutions that make Boulder unique: the sandstone-red buildings of the University of Colorado and its many high-tech labs, the campus of the Buddhist-inspired Naropa University, and the crystal stores and metaphysical bookshops that give Boulder its New Age flavor. These are the temples of Truth of our city: those places we go to seek a deeper understanding of reality.

Just underfoot is a long, low-slung building set against the foothills. This is NIST, the National Institute of Standards and Technology. NIST is the cathedral of physical science. In 2017, scientists here created what may be the coldest temperature in the Universe since the Big Bang, just .00036 degrees above absolute zero. NIST is also the home of the atomic clock, the most accurate time measurement tool ever created. In recent years scientists at NIST have led pioneering explorations into the deepest secrets of matter, devising ingenious experiments to separate sub-atomic particles but finding they're still linked somehow, soul mates tied together for eternity.

Naropa University, a small compound of brick buildings in the leafy center of Boulder, is one of the country's foremost Buddhist-inspired universities, founded in 1974 by the Tibetan teacher Chögyam Trungpa. Naropa is the home of truth-seeking through reflective practice and meditation. It's the place monks with shaved heads and saffron robes share classrooms with young women in dreadlocks and peasant dresses, plumbing the depths of contemporary Buddhist philosophy.

And dotted throughout the city are the white spires of churches, with several synagogues sprinkled in for good measure. These represent another form of truth-seeking, one grounded in faith, ritual, and tradition. These are the temples of monotheism, worshiping a set of beliefs passed down through three millennia and slowly shaped by the currents of culture and history to their present, Boulder-friendly form. I'm a sometimes-attendee of one of these, a family-friendly Reform synagogue in the middle of town called Har HaShem.

From way up here, these places all seem like dots on the same seamless actual-size map of Boulder, but I know that this is an illusion. The paths to truth represented by each of these temples are different, and often in conflict with each other. At NIST, physicists attempt to find a "unifying theory" to explain the ultimate physical reality of the Cosmos. At Naropa, seekers meditate in order to understand the Four Noble Truths, the succinct crystallization of Buddhist wisdom. And at Har HaShem, Jews are looking for Yahweh, the single source of all being.

I see value in each of these approaches to wisdom, but some disconnect nags at me. The stories told by each of these Temples of Truth seem contradictory. Conventional wisdom holds that ultimate truth lies either in scientific inquiry, or in religious insight, but not in both. Our culture teaches that rationality and spirituality are in conflict, and we must select one approach or the other. But how, I wonder, can there be multiple valid and opposing versions of ultimate truth? Either science and spirituality are in conflict – and one of them is *wrong* – or the whole idea that these two concepts are in opposition is drastically misguided.

I finish a well-deserved peanut butter cup, clamber off the rocks, and start heading down the Bear Canyon trail, a smorgasbord of wildflowers, aspen trees, and vistas north to Longs Peak and beyond. The gently meandering trail allows my mind to wander to a more meditative place. That nagging feeling is really a little voice calling out to me from the Universe – whatever *that* means – and telling me this: It's time for a quest. A seeker's quest, to pursue a unified understanding of reality.

My guiding principle on this quest will be to use science and reason as a lodestar, an intellectual rocketship to supercharge my understanding of myself and the Universe, and a filter through which to evaluate ideas from other seeker traditions. I've spent most of my professional life exploring the intricacies of the scientific method and the beauty and elegance of logic and mathematics, and I am convinced of their astounding power to explain reality. But on this journey I will also explore the limits of scientific thinking, and ponder what lies beyond that boundary.

And so I embark on a journey that will lead me from the ancient history of our species to the far reaches of twenty-first century science, from the great thinkers of ancient Greece to the philosophical implications of modern neuroscience and quantum theory, through Eastern mysticism and Western rationality. By the end of this amazing trip, I hope to arrive at a consistent, scientifically credible, and spiritually resonant story about the very nature of reality.

My love affair with science and reason begins as far back as my memory travels. As a ten-year old boy growing up in Massachusetts, I would have called myself a Jewish agnostic, but my primary spiritual identity was as a Red Sox fan. My connection to the game was not to play it, but to math about it, and my numeric obsessions found purchase tracking baseball statistics. Every afternoon, I'd cut the standings out of the *Springfield Daily News* and Scotch tape them with inky fingerprints to a sheet of white paper. I'd study batting averages and winning percentages, marveling at the way they rose (and more often, in those days, fell) as if conducting a secret mathematical dance.

Underneath this passion was a deep cynicism about any way of thinking about life that wasn't strictly logical. Being the smart, cynical rationalist wasn't great for dating girls or making friends, but it did make me *feel* smarter than the grown-ups around me who retained their obvious human frailties like religion and faith. This rationalist stripe eventually cohered into a deeply-felt atheism. It fit my contrarianism well, and seemed unassailably logical. The flaws of religion seemed so obvious that atheism was the only viable antidote.

I wobbled through college, where I proved to be a terrible student in every class except the one everyone else failed, statistics. After graduation, I remained mostly in academia for the next two decades, working as a staff assistant at Harvard University, dabbling in a few graduate statistics courses, and eventually pursuing a Ph.D. in education research here in Boulder at the University of Colorado. I loved the intricacies and hidden logic of linear regression formulas and statistical inference tests, but I also loved the philosophy of science class I took. The readings challenged the very assumptions about science I'd always taken for granted. Can we trust scientific results on their face, or are they filtered by our own preconceptions? Is there an ultimate truth out there, or is it all ultimately subjective? I gradually came to understand *why* science yielded such important truths about the world, and when it went awry. This was my mission trip, my indoctrination into the secrets of the code.

As analytic and eggheaded as my academic journey has been, I've always known that's not all I am. Those warm summer evenings spent huddling under my covers listening to Red Sox games were not only about numbers and math. When the game ended – typically in a lopsided Yankee victory – I'd turn off the radio and lie wide awake in bed. Our olive green three-family stucco house sat in a dense residential neighborhood, and outside my window the sky glowed orange from the sleepless lights of the city. I could always see one or two stars from my bed, and I'd study them, countless trillions of miles away. I knew there was something out there, something *big*, though I couldn't quite place my finger on exactly what. But I knew with surety I was part of it.

That *something* stuck with me as I navigated through college. After a long day studying math, I'd get drunk with my friends and then wander out alone into the forests behind my apartment, communing with the Berkshire night sky, listening for something. After being spit out into the Real World and enduring six years of existential gloom in slushy Boston, I packed up everything in my Subaru Outback and drove out to Colorado – not because of a rational algorithm, but because I simply intuited it was the right thing for me to do. I knew it was the right place. It was the story the Universe was laying out for me, and it was my job to write it.

Two decades later I'm still here in the mountains that have been calling my name my whole life. And right now, treading softly down the ridgeline, I feel an undeniable spiritual – even mystical – connection with his place, and with the very experience of being alive. There's some sort of truth there, I'm sure. How, exactly, that truth connects to the truth revealed by science is not exactly clear to me yet. But it's exactly what I hope to find out.

As I trot down Bear Canyon I stop for a moment at Saddle Rock, with its vista spanning from the Continental Divide across to the Great Plains. I scamper down in time to make it to a favorite restaurant for lunch.

Sherpa House is in an old Victorian mansion, slightly rejiggered into a cozy Tibetan restaurant. I'm led to a sunny table on the porch, surrounded by photos and newspaper clippings of the owner, Lhakpa Sherpa, and her extensive family. These folks truly know the meaning of a seeker's journey: the man cooking my lunch right this minute, Jangbu Sherpa, has summitted Everest ten times, and his family name is literally synonymous with the ultimate outdoor quest – climbing the highest mountain in the world. Sherpas are also members of the most ancient school of Buddhism, the Nyingma tradition. They place a high emphasis on *dzogchen*, described by the Buddhist teacher Sogyal Rinpoche as "the heart-essence of all spiritual paths and the summit of an individual's spiritual evolution." I'm struck by that beautiful correspondence: the physical quest to reach the ultimate external summit, and the spiritual quest to reach the ultimate internal one.

When my chicken *tikka masala* arrives, I dip into the tangy red sauce with a triangle of buttered *naan* and mull over the journey ahead. I've given myself a big hunk of a philosophical question to chew on: *What does scientific truth tell us about spiritual truth?*

A brief history of scientific seeking

I'm not, of course, the first person to explore this territory. Throughout history, people of all stripes, temperaments, and ideologies have trod the path of the seeker. Some are scientists who broke away from the cozy, safe assumptions of the day, like Galileo and Einstein. Some are religious mystical explorers, like the thirteenth-century German theologian Meister Eckhart or the twenty-first-century Tibetan monk the Dalai Lama. Others were solitaires and wanderers, like the nineteenth-century transcendentalists John Muir, Ralph Waldo Emerson, and Henry David Thoreau, who sought inspiration by escaping the corrupting influence of society to wander through their own philosophical forests.

Scientific seeking may seem a contradiction in terms to our modern way of thinking, but uniting the rational and spiritual realms hasn't always seemed so scandalous. In ancient Greece, for example, there was no meaningful distinction between "reason" and "God." Pythagoras, born on the island of Samos around 569 BCE, is mostly remembered today for his famous equation relating the lengths of the sides of a right triangle. But back in the day, he presided over a mystical cult that worshipped numbers. Pythagoras believed that mathematical rules describe everything in the Universe, and so understanding mathematics is key to unlocking the very structure of the cosmos. His cult saw his mathematical insights as mystical revelations.

The Stoics, established in Athens a few hundred years later, believed that reason is the very foundation of both humanity and the Universe. It followed that the goal of life is to live according to Reason, which was synonymous with Nature. Chrysippus, one of the most influential Stoics, wrote, "The Universe itself is God and the universal outpouring of its soul; it is this same world's guiding principle, operating in mind and reason." To the Neo-Platonists, who lived in the first half of the first millennium AD, the goal of human existence was henosis, meaning mystical oneness or union. The One is the source and end of all things, and the world is fundamentally rational and deterministic.

Over time, this mystical union between God and rationality fell victim to the growing power of the Church. By the middle of the next millennium, it was outright heresy to make claims based on scientific findings that contradicted the authority of the Church's teachings. Nicolaus Copernicus, born in 1473 in present-day Poland, foreshadowed this tension with his heliocentric model of the Universe, which placed the Sun, rather than the Earth, at its center.

Copernicus had a complicated relationship with the Church, and was an active and practicing Catholic. The friction between science and Church came to its apex fifty years later with an even bolder scientist, Galileo Galilei.

Galileo was born in 1564 in Pisa to a famous lutenist, Vincenzo Galilei. Galileo was a polymath, astonishingly capable not only in mathematics, but also in physics, astronomy, and music. Galileo's embrace of heliocentrism embarrassed the Pope, Urban VIII, who had urged Galileo to include counter-arguments to heliocentrism in his work. His ideas were controversial not only in the Church but in the scientific community as well. Among the eminent scientists who opposed the new heliocentric model were Tycho Brahe, an ardent believer in empirical evidence, who, incidentally, lost his nose in a duel with his cousin over who was the better mathematician. Galileo countered that his work was not in conflict with religion, recollecting his conversation with Cardinal Baronius: "The bible teaches us how to go to heaven, not how the heavens go."

Unfortunately, Galileo's defenses were not enough to counter the power of theocracy. He was found guilty of heresy in 1633 and was held under house arrest for the rest of his life, his books banned into the next century. But the revolution begun by Copernicus and amplified by Galileo would simply be too powerful – and reveal too much of the truth – to be defeated by the theocratic machine.

At the same time Galileo was battling the Church, the scientific revolution was blossoming in full. In the early part of the 17th century, the English philosopher and statesman Francis Bacon developed rules of inquiry that began to form the basis of the scientific method. Bacon, though, was no heathen. A devout Anglican, Bacon, like Galileo before him, saw his empirical philosophy as compatible with the religious teachings of the Church. "Knowledge is the rich storehouse for the glory of the Creator and the relief of man's estate," he said. For Bacon, while empirical observation and inductive reasoning were the best ways to learn about the natural world, knowledge of God could only be had through revelation.

As society liberalized, scientists got bolder in their outright defense of science against religion. In 1874, John William Draper, a philosopher, chemist, and early pioneer of the new technology of photography, articulated the "conflict thesis" of science and religion. "The history of Science," he wrote, "is not a mere record of isolated discoveries; it is a narrative of the conflict of two contending powers, the expansive force of the human intellect on one side, and the compression arising from traditionary faith and human interests on the other."

By 1879, Charles Darwin could declare himself an agnostic and still retain his status as a revered son of England, offering such iniquitous pronouncements like this one: "Science has nothing to do with Christ, except insofar as the habit of scientific research makes a man cautious in admitting evidence. For myself, I do not believe that there ever has been any revelation. As for a future life, every man must judge for himself between conflicting vague probabilities." A few years later, in 1896, Andrew Dickson White, the first president of Cornell, wrote, "In all modern history, interference with science in the supposed interest of religion, no matter how conscientious such interference may have been, has resulted in the direst evils both to religion and to science—and invariably. And, on the other hand, all untrammeled scientific investigation, no matter how dangerous to religion some of its stages may have seemed, for the time, to be, has invariably resulted in the highest good of religion and of science." The philosopher William James thought the spiritual realm is ultimately inaccessible to reach by rational means: "The attempt to demonstrate by purely intellectual processes the truth of the deliverances of direct religious experience is absolutely hopeless."

Over a century later, little progress has been made in uniting science and religion. Because their methods and goals are so different, many scientists today see them as fundamentally incompatible. Science is a rigorous inquiry into the true nature of the world and has the distinctive ability to revise itself as new evidence appears. Religion, on the other hand, holds that truths are unchangeable and are revealed not through rationality, but revelation. Stephen Hawking summed up the views of the scientific mainstream when he said, "There is a fundamental difference between religion, which is based on authority, [and] science, which is based on observation and reason. Science will win because it works."

A more charitable view is that science and religion are not in conflict, but are independent, with different "aims, objects, and methods." Stephen J. Gould wrote that science simply cannot comment on the existence of God. Science is meant to use empirical evidence and inductive reasoning to come to truths about the natural world, while religion takes on the big questions of meaning and God. For the Harvard University astronomer Owen Gingerich, a devout Mennonite, the ultimate question of why the Universe exists is "not for science to grapple with." Rather, it is a teleological question – that is, its answer must be an ultimate explanation or purpose, not a rational explanation. Or, as the French philosopher Blaise Pascal put it more poetically two centuries earlier: "The heart has its reasons which reason does not know."

Even mentioning "science" and "faith" in the same sentence is somewhat of a dog-whistle for atheists and rationalists. A case in point is a 2007 editorial in the *New York Times* by the physicist Paul Davies, which provoked a firestorm of reaction among the scientific community. In the column, titled "Taking Science on Faith," Davies argued that the idea that the laws of physics are immutable and fixed is an article of faith among physicists. Scientists investigate the effects of those laws – such as the speed of light – but take their existence and immutability as given. "To be a scientist, you had to have faith that the Universe is governed by dependable, immutable, absolute, universal, mathematical laws of an unspecified origin," Davies wrote. Both monotheistic religion and orthodox science fail to provide a complete account of physical existence, because both are founded on faith – for religion, the faith in an unexplained God; for science, in an unexplained set of physical laws. The idea that there are fixed, immutable laws governing the Universes stems from Christian theology, Davies argues.

On the website Edge.com, a variety of scientists expressed umbrage at Davies's idea. The basic gist of this criticism, expressed by Sean Carroll, is that there is no ultimate explanation for the

Universe, and none is needed. Moreover, it is foolish to pursue one: "That's just how things are." PZ Meyers, the biology professor and stridently atheist blogger, describes Davies as an "apologist for deism," assuming that this is the only conceivable reason the Times could have printed the article as "they certainly couldn't have chosen to print it on its merits."

The partisanship of this debate reveals a deep and unfortunate level of dogmatism in a community that prides itself on "freethinking." As Davies himself laments: "after 30 years of listening to sterile bickering in the science/religion debate I am utterly bored with the refrain from each side that, in effect, 'my superturtle is better than your superturtle.'"

In my own account, neither side is quite right or fully wrong. Religion and science are fundamentally different ways of approaching the truth, distinguished, essentially by definition, by whether they admit faith as a means to reaching truth. They cannot, however, be "separate but equal" paths to truth. If they both reveal ultimate truth, they must be fundamentally compatible. The revelations of one realm must be consistent with those of the other. Ultimate truth may extend *beyond* reason, but it may not *conflict* with reason. Similarly, reason may guide our journey to understanding, but it may not deny the immediate truth of mystical revelation – namely, the brute fact of self-awareness.

The science/spirituality debate can take on an academic, intellectualized flavor, but my journey, at heart, is a personal one. How do all these arguments and assertions translate into the inexpressible wonder of being alive? How do they help me understand the deeper purpose of my own life?

As I sort through the heady accounts of scientists and theologians and philosophers, I return to a book I haven't picked up in nearly three decades. Just out of college, at the very beginning of my own quest into adulthood, I received a copy of *Zen and the Art of Motorcycle Maintenance*. Published in in 1974 by Robert Pirsig, a

Montana State University English professor, the book had long since taken on legendary status when I came across it in the early 1990s.

Pirsig's story is a slightly fictionalized philosophical memoir that details his attempts to unite what he calls classical and romantic ways of understanding truth. His own journey lends the seeker's quest an appealingly romantic drama. "Few people travel here," he writes of what he calls the "high country of the mind." "There's no real profit to be made from wandering through it, yet like the high country of the material world all around us, it has its own austere beauty that to some people make the hardships of traveling through it seem worthwhile. Many trails through these high ranges have been made and forgotten since the beginning of time."

Picking up *Zen* after all these years, I feel a strong kinship with Pirsig's scientific seeker's quest. It's a masterpiece of memoir, a candid profile of a brilliant man battling the demons of mental illness and trying to find meaning in the ever-pounding drum beat of modern life. But there are some significant flaws I'd missed the first time around. For all his personal insights, Pirsig insists on operating in an intellectual vacuum. He occasionally calls on Western philosophers like Aristotle and Plato, but mostly to criticize them.

Alas, for all its kultur-bearing power, and its millions of copies sold, Zen and the Art is not taken seriously by modern philosophers. It's simply too contemptuous of the body of thought of others, too myopic in its own worldview. "Phaedrus," Pirsig writes, "following a path that to his knowledge had never been taken before in the history of Western thought, went straight between the horns of the subjectivity-objectivity dilemma and said quality is neither a part of mind, nor is it a part of matter. It is a *third* entity which is independent of the two." Only a brilliant narcissist could imagine that he was the first person in the history of Western thought to attempt to unite the physical and mental world in this way.

Still, there's something wonderful about the book. It's a celebration of the seeker's journey through untrammeled intellectual and spiritual terrain. That's a journey to embrace.

Eight rules for scientific seeking

As I set out on a journey of my own, I want to do my best to avoid the pitfalls others have encountered. So I've pulled together a set of rules to guide my own journey.

1. Embrace shoshin

In Zen Buddhism, the concept of *shoshin*, or beginners mind, means releasing your assumptions and preconceptions and opening yourself up to new ideas. "In the beginner's mind there are many possibilities, in the expert's mind there are few," said Zen teacher Shunryu Suzuki.

Shoshin requires a certain humility about one's understanding of the world. Our default ways of understanding the world are the joint product of our environment and our brains, themselves the product of millennia of evolutionary pressure. If science and reason reveal the truth of the Universe to be radically different from the story our culture and our brains tell, we will need to be bold enough to consider these truths, even (and especially) when they challenge and disturb our comfortably entrenched patterns of thinking.

2. Believe in truth

It seems obvious beyond stating that we ought to believe in truth. But there are many schools of thought on the topic of ontology – the study of what exists – and not all agree on the nature of truth.

Empiricists, historically represented by the philosophers John Locke and David Hume, assert that all knowledge comes to us through the outside world, via our senses. Locke's famous image of the "tabula rasa" – the blank slate – represents an empty mind ready to be filled with the experience of the world. Hume advised his colleagues to "hearken to no arguments but those which are derived from experience ... [to] reject every system of ethics, however subtle or ingenious, which is not founded on fact and observation."

Rationalists, on the other hand, believe that our knowledge comes through *a priori* understandings of the world – that is, through hardcore reasoning and logic. (The famous example of an *a priori* truth is "all bachelors are unmarried"). The rationalists, who count among their ranks Plato, Descartes, and Spinoza, believe that we can use our powers of deduction alone to understand how the world works without (in its extreme form) a shred of external evidence.

Though empiricists and rationalists seem to live at opposite ends of a spectrum, they have something in common: the belief that the truth *actually exists*. This lies in contrast to postmodernists, like the French philosopher Michel Foucault. "Truth is a thing of this world," he wrote. "It is produced only by multiple forms of constraint and that includes the regular effects of power." It's true that our *path* to truth can be affected by power structures; look no further than the "debate" on climate change to see how powerful energy companies with a vested interest in obscuring scientific facts have compromised the *transmission* of truth. But the idea that truth itself is just a product of power relationships is deeply cynical, not to mention highly anthropocentric – that is, it assumes that what goes on among human beings is so essential that it alone creates truth. Surely some objective reality existed before our particular egoistic species evolved to sit in Parisian cafes and write long tracts about it. As the philosopher Harry Frankfurt has said, postmodernists are "shameless antagonists of common sense." In order to seek the truth, we must believe that it exists.

An appropriately clunky old-fashioned word, alethiology, refers to this exploration of what truth actually is. Its clunkiness reminds us that we can get so caught up in thinking *about* how to find the truth, that we forget that there is, in fact, a truth out there that may not care whether we're looking for it or not.

3. *Eschew gurus*

Science is predicated on the idea that the truth is available to all of us, so long as we follow a few simple guidelines. We are all, in a sense, enlightened, in that we all have the capacity for rational inquiry, coupled with the gift of direct experience. For those many areas we cannot directly explore ourselves, we ought to rely on experts, not gurus.

Experts are regular human beings who have invested time and thought into the systematic inquiry of a particular topic. They critically evaluate what others have learned and come to understand and respect their arguments, even if they do not agree with them. Experts exist within a community of learners and seek to expand their own expertise by learning from others and modifying their own base of knowledge.

Gurus, on the other hand, are members of a supposedly enlightened class who claim they have a unique portal to direct understanding of the truth. Their membership in this class may be self-appointed – like Scientology founder L. Ron Hubbard – or it may be appointed by others, like the Dalai Lama or Jesus Christ. Gurus are retailers, middlemen (or women) who try to mediate other people's personal connection with the truth. When we exalt gurus, we foster the imbalances and bad behavior that occur whenever one person is vested with too much power. Look no further than Chögyam Trungpa, that revered founder of Naropa, to see how abuses of power– in his case, alcohol abuse and sexual predation – arise so easily in the spiritually anointed.

Wisdom comes from expertise and experience, not in God's capricious allotment of divinity to certain chosen ones.

4. Stand on the shoulders of giants

In 1675, the philosopher and polymath Robert Hooke wrote to Isaac Newton, requesting a private correspondence and maybe even collaboration with the great scientist. Newton wrote back, "I am not so much in love with philosophical productions but oft I can make them yield to equity and friendship … if I have seen further, it is by standing on the shoulders of giants."

One of the key attributes of a great scientist is a willingness to study the work of those before, and a respect for that earned expertise. Scientific findings may be attributed to individuals, and may in some sense arise through a burst of creative intuition, but in general it's a community effort. Even the most original thinkers drew upon the knowledge of their day. As revolutionary a breakthrough as it was, Einstein's theory of relativity drew heavily on the work of his peers, including Lorenz, Poincaré, and others. The great Newton himself spent hours at Trinity College studying the works of Aristotle, Descartes, and Galileo before developing his own mathematical innovations.

It's not enough to merely pay lip service to ideas that grab us, to pick out a pithy phrase or two to lend color and authority to our pronouncements. If we really wish to be good scientific seekers, we must pry a bit, delve through some primary sources, understand the nuts and bolts underneath the theory, watch the gears turning.

Respect and understanding for the deep thoughts of others is also an act of humility: it acknowledges that we have a debt to those who have brought our thinking to this point, and it acknowledges that there is far more we do *not* know than that which we know.

5. Be a healthy skeptic

The term "skeptic" comes from the Greek *skepsis*, meaning "inquiry." The original Skeptics were the followers Pyrrho of Elis back in the 4th century BCE Their goal was *ataraxia*, or peace of mind, which they achieved by pitting dogmatic ideas against each other until no knowledge was left at all. (Those ancients could be almost comically postmodern in their drive to attain *ataraxia*; the Sophist Cratylus believed that even words themselves are meaningless because they are fixed but refer to the ephemeral and changing world.)

It wasn't until the middle of the first millennium AD when skepticism began to take on its modern connotation, a reluctance to believe in dogmatic truths. Modern skepticism is not a worldview in itself, but a mindset, a process. A skeptic is a Bayesian – that is, she factors in prior assumptions about the world before drawing conclusions based on new information. A skeptic's approach to a new health claim, for example, would be to hold out a healthy suspicion of doubt until rigorous evidence has been provided. *Perhaps* a new nutritional supplement has some beneficial effect, but prior experience tells us most new "medical miracles" are anything but.

Skepticism ought to be distinguished from its cousin, cynicism. The term "cynicism" originated in ancient Greece, around the time of Socrates. The original Cynics believed the purpose of life was to live in virtue, and the best way to get there was to divest oneself of mortal pleasures, like sex, power, and even shelter – the famous Cynic Diogenes lived in a pot on the streets of Athens. In the modern sense, a cynic is someone with a negative outlook on life or human behavior. They believe that people are just out for their own interests. Cynicism is the self-perpetuating disease that infests our political system, amplified by attack ads and 24-hour news cycles.

Skepticism, unlike cynicism, is not in conflict with spirituality, nor does it imply nihilism or meaninglessness. A skeptic believes the truth is out there, and it very well may be profoundly different than his current understanding. As skeptics, we adopt the so-called Sagan standard, popularized by the late, great astronomer: "Extraordinary claims require extraordinary evidence."

6. Be aware of our own cognitive biases

Cognitive biases are mental processes that systematically cause us to believe irrational things. We all have them, simply by virtue of being human. Our brains evolved through evolutionary selection pressure. Our ancestors were more likely to survive and pass on their genes if their brains highlighted those things that were likely to be immediate threats, like a lion pacing in the tall grass. Quickly categorizing and reacting to those threats was essential, even if sometimes our immediate assumptions were wrong. If the lion turned out to be an orange rock, well, we got a little worked up over nothing – no big deal. But if it really was a lion, then panicking likely saved our lives, allowing us to convey our genes to the next generation.

Unfortunately, this evolutionary inheritance also makes us prone to superstition, logical fallacies, and racial biases. The list of known cognitive biases is long, including classics like the Fundamental Attribution Error, in which we tend to attribute people's innate personalities to their behavior rather than to external situational factors, and more arcane biases like the humor effect, whereby funnier things are better remembered than serious ones.

Consider this example, drawn from the experimental research of the cognitive scientists Daniel Kahneman and Amos Tversky:

> Imagine you are a physician working in an Asian village, and 600 people have come down with a life-threatening disease. Two possible treatments exist. If you choose treatment A, you will save exactly 200 people. If you choose treatment B, there is a one-third chance that you will save all 600 people, and a two-thirds chance you will save no one. Which treatment do you choose, A or B?

Almost three-quarters (72 percent) of respondents chose Treatment A, saving exactly 200 people. Next, respondents were shown a different scenario:

> You are a physician working in an Asian village, and 600 people have come down with a life-threatening disease. Two possible treatments exist. If you choose treatment C, exactly 400 people will die. If you choose treatment D, there is a one-third chance that no one will die, and a two-thirds chance that everyone will die. Which treatment do you choose, C or D?

In this case, most respondents (78 percent) chose treatment D, with a one-third chance that no one will die. But under close inspection, you'll see that these two scenarios actually offer exactly the same choices. Treatment A in the first scenario is identical to Treatment C in the second, and Treatment B is identical to Treatment D. Yet people make very different choices when the scenarios are worded differently. According to Kahneman and Tversky, this occurs because people evaluate gains and losses differently – people are more willing to take risks when it comes to losses, but prefer a "sure thing" when it comes to gains.

Our tendency to hold inconsistent or irrational beliefs is not always a bad thing. "Cognitive dissonance is often considered a failure of the human psyche," writes Yuval Noah Harari, the Israeli historian and philosopher and author of *Sapiens*. "In fact, it is a vital asset. Had people been unable to hold contradictory beliefs and values, it would probably have been impossible to establish and maintain any human culture." Our brains evolved in order to make quick decisions in times of danger, at the expense of occasionally lumping things incorrectly into the wrong category.

But now we know better. Our best weapon against cognitive bias is simply awareness. In the words of the Nobel Prize-winning psychologist Daniel Kahneman, "we can be blind to the obvious, and we are also blind to our blindness." Once we respect our own biases, we can take measures to counteract them.

7. Define Our Terms

Fuzzy thinking is a central hazard in the spiritual literature. Because we humans express our complex ideas through words, fuzzy thinking often arises through imprecise, inaccurate, or ambiguous use of language. We ought to do the best we can to clearly define the terms and concepts we employ. If we make up new terms, we ought to explain how they align with existing ideas and scientific concepts.

For example, consider the frequently misused and abused word "observer" in discussions of quantum theory. Scientists have found that determining the specific outcome of a quantum process is dependent on an outside "observer." This discovery is often misinterpreted to mean that consciousness is a required component of quantum mechanics. But an "observer," as physicists use the term, is not necessarily a conscious human, but rather any physical mechanism that interacts with the quantum system. In this case, an incorrect definition of this simple word quite literally changes the world.

Sometimes, of course, nonscientific metaphors can be useful to understand spiritual truths or express complex psychological experiences. For example, the writer Eckhart Tolle defines a "pain-body" as an "energy entity" that has its own "energy frequency." *Energy* does have a scientific definition: a fundamental, measurable component of nature that, when transferred to objects, grants them the capacity to do work. But that's not how Tolle uses the term. An "energy frequency," in the New Age sense, is a metaphor, not to be confused with any sort of physically measurable, precisely definable aspect of the natural world. In spite of this, Tolle's concept may still be useful; we just need to be clear that we are not referring to the scientific, literal meaning of the term.

8. Look Within

Finally, let's add one more component that's not on the usual list of scientific guidelines. It's what makes a *scientist's* quest a *seeker's* quest. We must be willing to look deep within ourselves, for that is where the heart of spirituality lies. To look within means to draw upon one's own experience, to follow one's nose.

Looking within requires us to develop two habits: *patience* and *silence*. Patience gives us the time and space we need to travel at our own pace and get into the weeds when we need to, to detour into areas that appear promisingly fecund, to stumble onto dead ends and be willing to turn back and carry on.

Silence – the process of shutting up and listening – gives us room to reflect and absorb these big ideas, giving the analytical part of our brains a chance to rest and recover. Our culture eschews solitude, and many of us have an internalized fear of being alone. Reaching the peak of our own spiritual summit is inherently solitary. It is, by definition, a place in which only we can stand. But when we do reach that summit, we connect to the greater truth we all share, and then we need not be lonely at all.

Our cosmic curiosity, our belief in wonder, our stretching to the divine, is the final, magic ingredient to be mixed into our scientific seeker's toolbox.

The road ahead

Seekers come in various types, but for purposes of convenience I will arrange them into three overlapping groups: rationalists, spiritual seekers, and the traditionally religious. If you're in any of these groups, I believe you'll find much here that's familiar, but also some ideas that will challenge your understanding. If we can get the scientists, the new agers, and the believers aligned in the same direction – in the direction of truth – we'll have immense power to change the world for good.

In many ways, I'd place myself in the first category. I hold a deeply rational worldview, and I believe in the power of the scientific process to uncover deeper truths about the world. I believe that climate change is real and is caused by humans and that vaccines are one of the greatest human advancements of the last century (and, it ought to go without saying, do not cause autism). I believe that the fact claims of religion are generally made up by humans, and that religion has often been used to reinforce power hierarchies and to divide groups rather than join them. I'm also a confirmed skeptic, and approach what I loosely categorize as New Age beliefs – from homeopathic cures to paranormal experiences – with a healthy dose of doubt. If this type of thinking resonates, then you'll relate to much of the flavor of this book.

But I'm not only a skeptic. I'm a spiritual seeker, too. Rationalism is often seen as synonymous with atheism, but there are many more philosophical options open to us. Accepting critical thinking, evidence, and reason does not mean throwing out the beating heart of existence. If concepts like *miracles,* or *energy,* or *manifesting,* or *universal consciousness* speak best to you, you might be a modern spiritual seeker too. If so, I think you'll connect to much of what's in these pages. If, in particular, you are drawn to princples of Eastern thought, especially the concept of non-duality, you'll find some uncanny overlaps with the modern scientific worldview. Not all precepts of Buddhism necessarily stand up to rational inquiry; it is a fixed belief system too, and in its classical form can be dogmatic and hierarchical. In true Buddhist form, we will adhere to the 9th century *koan* by the Zen master Linji Yixuan: "If you meet the Buddha, kill him." Our path is guided by the search for the truth, not an attempt to corroborate any existing belief system.

And finally, if you identify with a Western monotheistic religious tradition, you are about to join in on a journey that provides some credible scientific underpinnings for many tenets of monotheistic religion so often scoffed at by secular thinkers. Most strikingly, the idea of an all-powerful, all-knowing, benevolent God is in many ways consonant with a scientifically coherent worldview. But, religious believer, you too will be challenged by the demands of logic and rigorous scientific inquiry. A thoughtful philosophical exploration of divinity is different than uncritical acceptance of religious dogma, and many of the assumptions of monotheistic traditions will fall by the wayside. Still, we will find much common ground between theological and rationalist views.

As we embark on this journey together, I'll try to keep jargon and technical terms to a minimum. Books about science and philosophy are rife with complicated language intended to demonstrate the author's immense brainpower and/or remind his readers that they are part of a special, intellectually entitled cult. High-reaching theories of truth and spirituality – yes, this is one of those – often rely on complicated webs of invented terms, convoluted hierarchies, and page after page of indiscernible technical detail. That said, good reasoning often requires precision, so some multisyllabic terms may creep in here and there.

And lastly, a word on qualifications. By trade, I am an education researcher. While I do have a deep background in the process and philosophy of science itself, I am certainly not close to expert in many of the scientific domains I explore. No one could be. Scientific seeking is open to all of us, not just physics Ph.D.'s. In those areas in which I am not an expert, I have done my best to fairly reflect the scientific debate or consensus. When I have encountered a discrepancy between my interpretation of the physical world and the experts', I've generally assumed the experts are correct. I enthusiastically welcome corrections, refinements, and suggestions from those who may have a more intimate connection with any of the areas I cover.

My real qualification to write this book is not my academic background or my scientific expertise. It's the fact that I'm a human, and a seeker, like you. The journey is ours together.

CHAPTER 2

The Wise Universe

If the world cannot be rational in my sense, in the sense of unconditional surrender, I refuse to grant that it is rational at all. It is pure incoherence, a chaos, a nulliverse, to whose haphazard sway I will not truckle.

– William James, The Will to Believe: And Other Essays in Popular Philosophy

When I was in eleventh grade I took an introductory French class. But instead of spending class time learning French, instead I spent the semester writing every number between 1 and 10,000 in my notebook. In part, it was a passive-aggressive rebellion against discipline and hard work, which was undoubtedly tied in some way to some deeper psychology – the fear that my actual work would not live up to my high scores on standardized tests.

But why *numbers*? Why not scratch Van Halen logos into the desk or practice writing my name in graffiti font, like other kids did? Numbers were my refuge, my entrée into some deeper mystery of the Cosmos. Numbers revealed a fundamental, reassuring order underneath the confusing and ever-changing world of high school cliques and unfathomable girls.

There is, indeed, something miraculous in the very fact that the world is ordered. This property seems so obvious, so ingrained in our thinking, that to even mention it seems tautological, even nonsensical. But it is deeply profound, it underpins every aspect of our lives, and it permeates every belief system, from radical materialism to religious fundamentalism. For this reason, it has been given deep and even mystical significance by philosophers, scientists, and spiritual seekers alike.

The Greek philosopher Heraclitus (c. 525 – c. 475 BCE) called the principle of order and knowledge *logos*. Philo (50 AD) saw logos, the divine Reason, as a mediating principle between matter and God, "the One and the All." This idea is reflected in the concept of natural law, that we can use rational thought to understand the fundamental rights and ethical responsibilities endowed by nature. Aristotle saw this universal law as "a natural justice and injustice that is binding on all men, even on those who have no association or covenant with each other."

We're getting a whiff of what Immanuel Kant, writing in Germany in the mid-18th century, called the Categorical Imperative. This principle holds that rationality is the ultimate good and people are fundamentally rational beings, and that universal moral laws must be built upon this rational foundation.

Perhaps no philosopher has embraced the tie between rationality and ultimate truth more profoundly than Baruch Spinoza, the seventeenth-century Jewish philosopher whom Bertrand Russell called the "noblest and most lovable of the great philosophers." Spinoza believed rationality defined God's ultimate nature, and that there was a deep, mystical tie between rationality and love. "Love," he wrote, "is nothing but Joy with the accompanying idea of an external cause." We cannot help but love the truth, because we need the truth in order to find our true selves and attain our purpose. A person who despises truth, therefore, must be someone who despises his own life.

This simple fact – that the Universe is ordered – is responsible for the "unreasonable effectiveness" of science and math, in the words of the physicist Eugene Wigner. The motivation for empirical science, writes Paul Davies, is that the world has contingency – that is, one thing causes the next in a predictable way. Davies calls this property "the dependability of nature." It is what allows inductive reasoning to be so successful. "It is this combination of contingence, rationality, freedom, and stability of the Universe," said Scottish Protestant theologian T.F. Torrance, "which gives it its remarkable character, and which makes scientific exploration of the Universe not only possible for us but incumbent upon us."

Interestingly, it may be impossible for the world to not be ordered, even in principle. What would such a world look like? We might imagine a universe with no patterns or laws at all, just an undefined cloud of *stuff*. If there truly are no laws governing its behavior, that stuff must be distributed randomly. But randomness itself has a sort of sense to it. For example, the moment at which a given atom in a radioactive element decays is completely random, and fundamentally unpredictable even in principle. And yet, we can predict exactly how long it will take for *half* of any given collection of atoms to break down – we just don't know which half. Even in its seemingly senseless randomness, the element's behavior seems to reflect some underlying order.

Rationality is the cognitive expression of order. And, indeed, it's almost impossible to *think* without rationality. That's because language itself is an inherently ordered process. Spiritual and philosophical statements, whether formal arguments or airy-fairy new-age proclamations, reflect a certain logical consistency. Even faith itself is underlain by a rational argument. For example, consider this passage from the Book of Chronicles:

> If my people, who are called by my name, will humble themselves and pray and seek my face and turn from their wicked ways, then I will hear from heaven, and I will forgive their sin and will heal their land. Now my eyes will be open and my ears attentive to the prayers offered in this place.

And now take a look at this passage, from the blog *Energy Muse*:

> Chakras are energy centers along your spine. There are seven main chakras, known as the root, sacral, solar plexus, heart, throat, third eye and crown chakras. … Each of the 7 chakras has a vibrational frequency that is associated with a different color. In energy healing, including crystal therapy, the chakra's meaning is linked with a particular set of emotional, physical, and spiritual issues. When you pair certain crystals with chakras in the body, the crystalline structure of gemstones work to amplify your healing intentions and restore and rebalance the energy body by removing blockages.

Neither of these passages comes from a scientific or rationalist worldview. The first proclaims that if you pray to the monotheistic God of the Hebrew Bible He will forgive your sins and heal your land. The second states that the crystalline structure of specific gemstones restores energy within each of the body's seven chakras. Both of these statements use logical reasoning, drawing upon a presumed order in the Cosmos. *If* you pray to God, *then* He will heal your land. *If* you use the right crystal, *then* its internal structure will rebalance your energy.

Of course, using a logical structure does not mean the statement is valid. Praying to the right god will not make your crops grow any more than crystalline structures in gemstones will heal your broken bone. The point is that even those allegedly operating on a plane above rationality must resort to logical thinking to make their case, even if that logic is flawed.

Math is nature

A prevailing academic truth throughout college was that everyone hated a singular course, usually labeled "sadistics." I loved it. Statistics was orderly and intuitive and whispered truths about an underlying cosmic structure that remained invisible to the untutored.

The observation that mathematics illustrates a profound truth of the Universe dates back at least 2,500 years, when Pythagoras and his cult followed the doctrine "all is number." Scientists ever since have used math to reveal an eerily profound wisdom. As Galileo said, "the book of nature is written in mathematical language." Or, in the words of the physicist Paul Davies, "No one who is closed off from mathematics can ever grasp the full significance of the natural order that is woven so deeply into the fabric of physical reality." And Eugene Wigner, the Nobel-prize winning physicist, wrote, "The miracle of the appropriateness of the language of mathematics for the formulation of the laws of physics is a wonderful gift which we neither understand nor observe."

A modern example of Pythagoras's doctrine comes from the field of information theory. Many contemporary mathematicians believe that the Universe is ultimately composed of perfect and timeless mathematical objects, such as pi or the square root of -1. The universe can be seen as a huge computer that holds these objects. (Interestingly, this would imply that mathematics was meaningless shortly after the Big Bang, because particles in the early computer-as-universe were too far apart to meaningfully communicate, and thus compute.)

In my own realm of educational research methods, I've been struck by how those dry old statistical techniques actually seem to provide glimpses into an underlying cosmic order, once we grok their intrinsic logic. A t-test may have been invented by a scientist to solve a practical real-life problem (in this case, to monitor the quality of Guinness stout), but it works because it speaks the hidden language of the Cosmos.

Not only is the Universe ruled by mathematics, but we just happen to have minds that can understand it. David Deutsch, an Israeli-British physicist at Oxford University, calls this the principle of comprehensibility. "There's a quasar out there in space. Billions of light-years away. And in our brain there's a model of the quasar, a model that has remarkable properties. There's not just an image of the quasar in our brain, there's a structural model with the same causal and mathematical relationships."

The explanatory power of mathematics is clearly significant. But where do these mathematical ideas come from in the first place? The two major schools of thought on the topic go back to Plato and his student, Aristotle. Plato believed in the ever-changing physical world on one hand, and an immutable, eternal world of Ideas on the other. Pure mathematics lives in this Idea world. Aristotle, on the other hand, believed mathematics is created, not discovered, built through a process of logical operations. Mathematics is a verb, not a noun – a formal procedure rather than something that exists in its own right. This school of thought, therefore, is known as *formalism*.

Many physicists, such as Max Tegmark and Roger Penrose, are Platonists. They believe every possible mathematical form exists, for real, out there somewhere. Tegmark takes the radical view that each of these mathematical structures constitutes its own parallel world, and together they make up a mathematical multiverse. To many mathematicians, these structures have a certain inherent beauty which guides discovery. For example, the theoretical physicist Paul Dirac used aesthetics as a guide to improve his equation for the electron. "It is more important," he said, "to have beauty in one's equations than to have them fit experiment." Paul Davies believes that the laws of physics have their own, independent existence, and are not merely beliefs that reside within people's minds and are created by them.

So which is it – is math really *out there*, or is it just something we *do* to better understand reality? Two and a half millennia after the ancient Greeks first raised the issue, the issue remains unresolved. Mathematicians, it is said, are Platonists on weekdays and formalists on weekends. When they are in their labs and working through equations in their offices, the sense of discovering connections and patterns in math truly feels like uncovering truths that have some independent existence of their own, just waiting to be found. Then, when they have a chance to go home, have a beer, and mull over the philosophical implications of their findings, their week's toil seems more like a game, a process of using made-up symbols in a consistent way.

My own journey as a scientific seeker has given me a great deal of empathy for the plight of mathematicians caught between the Platonist and formalist ideas. I have gained a deeper appreciation of the order underlying reality, the recurring patterns and elegant equations that describe our world with remarkable accuracy. But I've also seen that the way we understand the world is really a story created by our brains. It is *interpretation* of brute reality, which itself is beyond descriptions and definitions. Reality just *is*.

Whether true or not, the Platonist model doesn't help us answer the deeper philosophical and spiritual questions of the Universe. As Stephen Hawking famously wondered, "Why does the Universe go to all the bother of existing? What is it that breathes fire into the equations and makes a universe for them to describe?"

The scientific approach of constructing a mathematical model to describe reality cannot answer why there should be a universe for the model to describe. It's as if, as the physicist John Archibald Wheeler suggested, we were to write down all the laws of physics on bits of paper and scatter them over the floor, then stand back and say, "Fly." They won't fly. They just sit there.

Mathematics is part of a larger ordered truth. We might call this higher-order truth *wisdom*. We live, then, in a wise universe – one whose totality is beyond our comprehension, but is infused with the miraculous property of order. It is this underlying wisdom that allows the natural world to exist, and allows us to be here to experience it.

The birth of science

Over the past few centuries, humans have devised a tool to take advantage of the Universe's order and plumb its secrets for our own benefit. We call this tool *science*.

The principles of systematic inquiry were not delivered fully formed to the human race, but emerged over time, gradually becoming more defined. Ibn al-Haytham was one of the first to sketch the outlines of the scientific method, back in 1010. Though a devout Muslim, he was a scientific skeptic. In his *Doubts Concerning Ptolemy*, Ibn Al-Haytham argues, "Truth is sought for its own sake. And those who are engaged upon the quest for anything for its own sake are not interested in other things. Finding the truth is difficult, and the road to it is rough."

Science emerged as a formal field of thought around the time of the enlightenment, in 16th century Europe. Francis Bacon was one of the first to formalize rules for scientific inquiry. "The human understanding is like a false mirror," Bacon wrote, "which, receiving rays irregularly, distorts and discolors the nature of things by mingling its own nature with it." His four "idols" identified many of the frailties of human thought we now call cognitive biases. Identifying and combating these biases, you'll recall, are one of the essential tools for scientific seeking.

- The "idols of the tribe" are our human tendencies to make generalizations, to exaggerate, and to extend our observations about the world we can see, to the world beyond. We observe the stars and attach a presumption of order to the cosmos.
- The "idols of the cave" are the thoughts which roam about our mind (the cave) and color how we see the world. The chemist sees chemistry in all things, the courtier immersed in the royal court overemphasizes the significance of the king.
- The "idols of the theater" are the preconceptions and assumptions built into the fields of theology, philosophy, and science that shape our thinking, preventing us from seeing what is really there.
- The "idols of the marketplace" are the way language itself, traded publicly in the currency of words, can be misleading.

Bacon's idols demonstrate a clear-eyed and forward-thinking understanding of the way in which our own preconceptions, and those of the culture we live in, compromise our search towards an authentic truth. For this reason, E.O. Wilson crowned Bacon the "father of innovation": "among the Enlightenment founders, his spirit is the one that most endures. It informs us across four centuries that we must understand nature both around us and within ourselves, in order to set humanity on the course of self-improvement."

Though Bacon was a devout Anglican, the scientific revolution he helped launch created an irreversible schism between science and religion. By the middle of the 17th century, "the mystical" was increasingly applied only to the religious realm, while "natural philosophy" had its own methods to discover the hidden meaning of the Universe. The term "scientist" finally emerged in 1834, when the Industrial Revolution was in full swing.

The twentieth century brought with it more revolutionary shifts in our understanding of the natural world. Einstein's theory of relativity revealed the mind-boggling truth that time is intimately connected to space and changes relative to the observer. Quantum physics overturned the Newtonian view of a mechanistic cause-and-effect world, supplanting that centuries-old billiards-table model with the radical insight that at the smallest level, the physical world is truly random.

The scientists behind the radical paradigm shifts of the past hundred years were not met with the same theocratic opposition as their forebears. The scientists at the heart of these revelations lived primarily in Western Europe and the United States, secular societies in which the all-powerful influence of the Church had waned. Science had gained the upper hand in its power to describe reality, even though religious fundamentalists continued – and still continue – to fight the now centuries-old scientific revolution at every turn.

As science emerged as a formal field of study, philosophers began to ask more incisive questions about its inner workings, the source of its powers, and its failings. The prevailing school of thought at the dawn of the twentieth century was logical positivism. This doctrine holds that the only meaningful scientific statements are those verifiable or falsifiable by empirical observation. The great philosopher of science Karl Popper believed that a scientific statement "must be capable of conflicting with possible, or conceivable observations." For example, the statement "nothing can go faster than the speed of light" is scientific, as evidence could, in theory, be found to prove it false. But a statement like "aliens landed at Area 51 but they are impossible to detect" is unfalsifiable, and therefore unscientific: given the way the claim is formulated, it is impossible to prove it false.

Using falsifiability as the sole criterion for determining what statements are scientific is problematic, however. Not all unfalsifiable statements are fallacious, or even wholly outside the realm of science. For example, some cosmologists have proposed the idea of a Multiverse: innumerable, wholly independent universes that together form all of the Cosmos. We can never communicate with – or even detect – these other universes, even in principle. According to Popper's criterion, the multiverse theory is unscientific. This seems unduly constraining, as the theory itself derives solely from the domain of science.

Published in 1962, Thomas Kuhn's *Structure of Scientific Revolutions* offered a radical new way of understanding scientific progress. Kuhn argued that science proceeds not in a steady, linear way, but in bursts. In these revolutions, which Kuhn called *paradigm shifts*, assumptions within a field are completely overturned and replaced with new ones. Well-known paradigm shifts include the shift from geocentric to heliocentric models of the Solar System, ushered in by Copernicus and Galileo; Einstein's relativity, which overturned assumptions about time and space; and quantum theory, which supplanted the Newtonian understanding of causality.

Although the theory of paradigms was revolutionary in itself – a paradigm shift in the philosophy of science, you might say – it did not challenge the idea that the truth is really out there, somewhere. Einstein did not *change* the actual truth of the Universe with his new theory; his work merely brought us a better understanding of it. It was the postmodernists, like the aforementioned Foucault and the twentieth century philosopher of science Paul Feyerabend, who questioned whether an absolute truth exists at all. Feyerabend saw scientific theories as incoherent and other forms of truth-seeking, like religion, as equally valid. In his view, "science should be taught as one view among many and not as the one and only road to truth and reality." Some postmodernists have denied the existence of an external reality at all. The English philosopher George Berkeley wrote that "the only things we perceive are our perceptions." To Berkeley, the material world outside our senses simply doesn't exist.

These postmodernists were rightfully criticized in the scientific community, accused of anti-intellectualism. However well-intentioned, the shadow of postmodernism can be seen today in the fake-news, Trump-esque view of facts as matters of opinion to be chosen for strategic or personal reasons, rather as objective, shared truths.

Refuting the idea that there is no external truth seems so obvious as to be self-evident and need no defense. Nevertheless, a school called empirical realism emerged to reaffirm that there is, in fact, a real world out there, and that we can objectively know it through scientific testing and observation. This is sometimes called the correspondence theory of truth, because it assumes a correspondence between the real world and our mental understanding of that world. Among its adherents are biologist Francis Crick ("there is an outside world ... largely independent of our observing it"), physicist Steven Weinberg ("the working philosophy of most scientists is that it is an objective reality"), and biologist E.O. Wilson ("modern progress is based on the Enlightenment principles of 'objective truth based on scientific understanding'").

The moral: the truth is our anchor, our lodestar. It is non-partisan, non-ideological, and apolitical. If we wish to find it, we ought to be aware of the societal dynamics, personal biases, and cognitive or perceptual limitations that make seeking it more challenging. But we cannot, for a moment, lose faith in its very existence.

The special power of science

Science has an essential property that distinguishes it from other truth-seeking paths: it changes. Science has the unique ability to revise itself in the face of new evidence. This, in turn, allows scientific knowledge to adjust itself in order to ensure consistency with the outside world. We can illuminate this property by examining the weaknesses of a belief systems that lacks it: religion.

Religious believers argue that their belief systems are a separate-but-equal, or perhaps superior, form of truth-seeking. These systems come with their own logic and rules. In Islam, Judaism, and Christianity, for example, the Ten Commandments are the operating rules of life, delivered by God. Break them – say, sleep with your neighbor's wife – and thou shalt be smitten by God. To a biblical literalist, these simple rules seem far more straightforward than the complex rules given to us by science, which tells us that energy is somehow equal to time; that a complex web of inscrutable calculations can somehow explain something as erratic as the weather; and that human behavior itself is not subject to clear rules but has arbitrary and ever-shifting moral consequences.

While at first glance religion and other unscientific belief systems seem to possess a consistency lacking in science, in fact the opposite is true. True, religion can achieve a sort of *internal* consistency, propping itself up by its own rules, but in general religions need not adhere even to this basic property. At the risk of being pedantic, we can see that the Bible itself is notoriously inconsistent, whether in the size of David's army (1 Chronicles 21:5 tells us 1,100,000, while 2 Samuel 24:9 tells us 800,000), or the manner in which God creates the world, whether by making, blessing, placing, or simply speaking. It was our friend Spinoza who called the Bible "in parts imperfect, corrupt, erroneous, and inconsistent with itself, and that we possess but fragments of it." His reward was expulsion from his religious congregation by the great rabbis who were too threatened by this truth to keep him around.

Theologians have defended Biblical inconsistencies by appealing to broader principles or loopholes, such as the idea that the real problem is not the Bible, but us. The apparent inconsistencies are due to incorrect interpretations and incomplete understanding of the historical context; the original text itself, once discovered, is inerrant.

But whether the Bible – or any other belief system – is internally consistent is a bit of a red herring. The more important question is whether it is *externally* consistent – that is, does it comport with our empirical observations of the real world. The answer is that religion can't possibly be externally consistent, even in principle. It is, by definition, a fixed set of beliefs, based ultimately on revelation and doctrine rather than on reason and empirical observation. For that reason, it can never keep up with our ever more sophisticated understanding of the natural world.

We often think of science as revealing truths asserted by humans, and of religion as revealing God's truth – but in fact, the truth is just the reverse. Science is our way of uncovering the external truth of the Cosmos, while religion is a human-created set of stories and arguments to help humans better cope with a mysterious and often hostile world. The power of science to reveal the true nature of the world is evident in the uncountable wonders it has given us, from modern medicine to the Internet to space travel. Over countless millennia, religion has never been able to cure polio, to bring humans to the moon, or to allow us to instantly communicate with our fellow humans on the other side of the world. In less than a century, science has achieved all of these miracles.

Science hasn't just given us technological achievements. It has also morally advanced our species. Among the most articulate proponents of this argument is Michael Shermer. In *The Moral Arc*, Shermer argues that as humans have become less religious, they have actually become *more* moral. "It is no longer acceptable to simply *assert* your moral beliefs. You need to provide *reasons* for them, and those reasons had better be grounded in rational arguments and empirical evidence or else they will likely be ignored or rejected."

Science has helped us understand that poverty is not a result of the inherent inferiority of particular groups of people, but due to comprehensible political and economic processes. The legalization of gay marriage owes credit to the supplanting of the traditional religious understanding of human sexuality by a scientific one, with morally beneficial consequences as a result.

If science and reason really improve our lives, we'd expect to see evidence that human life has gradually improved as science has had an increasingly large impact on our society. And that's exactly what we find.

Steven Pinker, the MIT psychologist and author, is one of the foremost proponents that the world is getting better over time, in measurable ways. He points out that the annals of history are littered with violent deaths. A man living in the prehistoric era – about 9,000 years ago – had a one-in-four chance of dying violently. Hunter-gatherer societies in South America, Australia, and North America lost, on average, nearly 20 percent of their populations to warfare. "Violent deaths of all kinds have declined, from around 500 per 100,000 people per year in pre-state societies to around 50 in the Middle Ages, to around 6 to 8 today worldwide, and fewer than 1 in most of Europe," Pinker writes.

While we seem to live in a violent, war-torn era, statistics reveal a somewhat different picture. In the twentieth century, only about 3 percent of people died from wars and war-related deaths like disease, famine, and genocide. While that death toll – 180 million souls – is undeniably tragic, it pales in comparison to the human-caused violence which came before. In societies that existed before the emergence of states, 15 percent of people were killed in wars. That would be equivalent to the violent deaths of over one *billion* people today.

The idea that traditional, pre-industrial societies somehow lived in greater harmony with nature is also a modern trope. The ecologist Bobbi Low analyzed 186 of these traditional societies and found that not only were they not practicing conservation, but they caused severe deleterious effects on the environment.

There is a strong argument to be made that the much of this improvement in the human condition is due to the progress of science and reason. That's because science is the best method we know of to get closer to the truth, so scientifically-oriented societies – those with a deep respect and understanding for science – will get closer to the truth. And as we get closer to the truth, we orient our society in ways that reflect truth.

The coronavirus pandemic, deadly as it is, illustrates this principle. As of January 2021, it had killed about 2 million people – or about 0.03% of the world's population. While this toll is extraordinary, it pales in comparison to previous pandemics, such as the Spanish Flu of 1918, in which 3 to 5 percent of the world's population died. Worse still, the Black Plague wiped out up to a third of the entire population of Europe. What distinguishes the coronavirus pandemic from previous ones is our understanding of the biology of the disease, enabling improved medical treatments to reduce its worst effects and the record-setting development of an effective vaccine, which will likely save millions upon millions of additional lives. All thanks to science.

Believing that the world is good, and that we can use science and reason to make it better, does not have to mean we must expect a magical, ideal future. Rather than a utopian, I might call myself a *protopian*, a term coined by the visionary futurist Kevin Kelly: "I believe in progress in an incremental way where every year it's better than the year before but not by very much – just a micro amount." Michael Shermer describes protopian thinking this way: "its prescriptions are modest and the general principle is relatively simple: try to make the world a slightly better place tomorrow than yesterday."

Bad science

What, you may ask, about science's dark side? What about its sinister social history, like the nineteenth-century phrenologists who "scientifically" studied skull patterns to conclude that Europeans were smarter than other races? Or the way science made possible horrific inventions of human destruction, using mathematics to create effective catapults in ancient Greece and particle physics to create the atom bomb in the twentieth century? And what about the many "discoveries" made in the name of science that have since been thoroughly debunked, from the use of bloodletting to cure heartbreak to the supposed link between vaccines and autism?

There are essentially two categories of complaints embedded here. The first criticism is that claims made by science turned out to be wrong, and not reflect the truth at all. The second is that science and technology has been used for destructive ends. Let's take a look at each of these challenges individually.

Yoshihiro Soto was a Japanese bone researcher who published extensively on his research into how to reduce bone fractures. His groundbreaking findings, documented in over 200 published research papers, were widely cited. One found that women who have had a stroke reduced their rate of bone fractures by a breathtaking 86% by taking a drug called risedronate. Other Sato studies investigated reductions in hip fractures via sunlight, vitamin D, vitamin K, and folate. All but two reported similarly dramatic effects.

Sato's results had a ripple effect throughout the medical community, affecting health guidelines and treatment practices. Unfortunately, he was a fraud. His data were fabricated, his findings falsified. His legacy represents one of the greatest failures in the annals of academic research to identify fraudulent findings.

The Sato case illuminates the obvious truth that science is done by humans, and it can be done poorly. But this type of bad science is not due to the limitations of the scientific process itself, but rather to humans' failure to perform science correctly. While Sato's actions were purposeful and nefarious, well-intentioned scientists have also arrived at results are that are misleading or just plain wrong. In one splashy analysis, the meta-science researcher John Ioannidis concluded that less than half of the findings reported under current publication guidelines are likely to be true. There are a host of reasons for this, such as "p-hacking," the tendency of researchers to selectively publish individual findings that appear statistically significant (i.e., they have low "p-values"), while consciously or subconsciously sweeping non-significant findings discreetly under the bed. Other studies have found that innovative findings tend to wear off over time, even when careful publication guidelines have been followed.

Some claim that all of these examples indicate that science itself is at some level untrustworthy, or, in the words of Paul Feyerabend, that "science prevails not because of its comparative merits, but because the show has been rigged in its favour... It reigns supreme because some past successes have led to institutional measures (education; role of experts; role of power groups such as the AMA) that prevent a comeback of the rivals."

What Feyerabend fails to distinguish is *science* as a profession conducted by humans; and *science* as a logical process based on the twin pillars of logical reasoning and empirical observation. The business of science can fall prey to financial incentives, publication bias, p-hacking, or Sato's fraud. But these problems don't indict science itself; they indict the people and institutions who compromise the scientific method. In other words, to get to the truth, we ought to peel away those things that compromise scientific inquiry, not throw out scientific inquiry altogether.

As an analogy, imagine you live in the year 1290, just before the invention of glasses. A time-travelling huckster floats into your town claiming knowledge of an amazing new invention that allows the previously blurry-eyed to see clearly! The huckster collects your few shillings and then half-heartedly grinds an ill-fitting set of lenses that barely work. Glasses, you say, are a fraud! They are no better, or no worse, than other means of being able to see, like divine revelation or dreams. The "science" of glassmaking is a societal institution subject to perverse incentives and power dynamics, you say, and can do no better or worse than revelation.

The truth, of course, is that the problem lies in the glassmaker, not the glasses. Glasses are better than revelation at seeing the world around them, but they still may be made poorly. Similarly, science is better at understanding the world than other methods, but it can still be done poorly.

Science is an imperfect process. It is an ungainly and often awkward animal. It clambers across time, intaking new research studies and investigations, some brilliant, some fatally flawed, most somewhere in between. Each new finding is a tiny piece that contributes to a growing puzzle. Sometimes the piece extends or emphasizes what we already know; sometimes it moves another piece to a new location or removes it altogether. Eventually, a macro object begins to take shape: knowledge. It is the accumulation of many inquiries, many revisions and responses, honed over time like a river stone.

There's a second criticism of science: it's dangerous not because it finds the wrong answers, but because it finds the right ones, which inevitably end up causing harm rather than good. Science and technology, goes the argument, are invariably used to dire ends, such as to drive warfare.

The rejoinder to this argument is similar to the one laid out above: science is an incredibly powerful tool that allows us to come to greater understanding of the world, whether that understanding is about the workings of the human body or the nature of the atom. It is a tool that can be used for good or evil. Science itself has no moral valence; only scientists do.

Bounds of rationality

Science is powerful. But is it powerful enough to explain *everything?* Some say yes: there is no bound beyond which science cannot venture. Science, they say, can explain everything, from the motion of the planets to the workings of the brain to the essential questions of existence itself, like the meaning of consciousness and the ultimate purpose of the Universe. This viewpoint is derisively referred to as *scientism*, which holds that science is the *only* reliable means of inquiry into the ultimate questions of philosophy and nature.

Most scientists do not adhere to the extreme claims of scientism; they know that science and rationality have their bounds. The question is in determining where, exactly, they lie. Popper called this process of separating the scientific from the unscientific "the problem of demarcation." He used falsifiability as his demarcation criterion, but there are many others.

Many modern philosophers see certain metaphysical questions – such as what happened before the beginning of the Universe – as beyond the boundary of science to understand. Alan Turing, the British mathematician who was instrumental in developing the conceptual underpinnings of the computer, left a "message from the unseen" behind at his death: "Science is a differential equation. Religion is a boundary condition." In his own highly nerdy language, he meant that science progresses, but it does so within the friendly confines of a defined equation. Religion is what is beyond the limits of the equation itself.

While science and reason can't fully answer all meaningful questions, they can take us further into the realms of the spiritual than you may realize. As an example, let's take one of those boundary-defining questions, typically thought to be beyond the demarcation line: Is there such thing as a soul?

The typical theistic answer to this question is that of course the soul exists, but proving its existence is beyond the capabilities of science. Atheists, on the other hand, place the question on the near side of the demarcation line, within the realm of scientific inquiry. For example, Yuval Noah Harari argues that because science has yet to prove the existence of a soul, it must not exist – and thus, there is nothing particularly special about humans.

Let's, instead, investigate a third possibility: that the soul does exist, and science can offer at least some evidentiary insight. The logical argument might go something like this:

1. By the "Soul," we are referring to its dictionary definition: "the part of you that consists of your mind, character, thoughts, and feelings."
2. Science works by making hypotheses about the world and then testing them through logical deduction and/or empirical observation.
3. The existence of our mind, and its thoughts and feelings, are provable by direct empirical observation. To paraphrase Descartes's famous *Cogito Ergo Sum* (I think, therefore I am), awareness implies existence.
4. The existence of the Soul is provable by direct empirical observation, which is a component of scientific inquiry.

In an admittedly narrow way, we've "proven" the existence of the Soul through science. But using logical arguments to demonstrate that something like the soul lies within the logical world leaves us with a new problem: we are using rationality itself to prove that the world must be rational. The argument eats its own tail. We are demonstrating the ultimate sovereignty of the method by using the method itself. As the philosopher of religion Keith Ward observes, a statement like "no statements are true unless they can be proven scientifically" cannot itself be proven scientifically.

The inescapable truth is that the power of reason must call upon itself as its primary justification. The mathematician Kurt Gödel identified this problem in his so-called incompleteness theorem. In a nutshell, this says that while the theorems of a system can be deduced from axioms, the axioms themselves cannot be justified from within the system. In other words, a given system (like mathematics, or the Universe itself) has essential truths (axioms), which can be deduced logically through a set of rules (theorems). A system may be internally consistent, but you can't use its rules to prove anything *outside* of the system.

If we apply Gödel's theorem to the system of rationality itself, we find that we can not use rationality to prove the validity of the *system* of rationality. We can never prove that the essential nature of reality is fundamentally rational. Perhaps the idea that we can come to fully understand the world through reason and empirical observation is an illusion in itself, an evolutionary adaptation that helped us persevere through the millennia. As Daniel Kahneman writes, "Our comforting conviction that the world makes sense rests on a secure foundation: our almost unlimited ability to ignore our ignorance."

But have you spotted yet another layer of irony as we've plumbed the story deeper? Gödel's theorem itself, and Kahneman's observation, are all *rational*. We just can't escape it as long as we're using words and logical arguments.

The best way I know of to escape this death spiral is to step outside of the whole discussion altogether. This approach is best captured in the form of Zen *koans*, such as the famous *koan* by the Zen Master Hakuin Ekaku:

> Two hands clap and there is a sound. What is the sound of one hand?

Or this one:

> A student asked Master Tozan, "What is Buddha?"
> Tozan replied, "Three pounds of flax."

Perhaps it's true that the ultimate truth of the Universe transcends reason itself. Importantly, though, this does not imply that the Universe is *irrational*. Ultimate truth may supersede rationality, but does not run counter to it, in the same way that Newtonian physics is correct but its domain is limited.

In spite of its limitations, it's quite remarkable the extent to which science has been able to explain reality. From the behavior of subatomic particles to the formation of galaxies, we've been able to put together a coherent picture of what we call the "physical" world. While this picture is by no means complete, it is remarkably good at explaining our observations of reality. This is the basis for naturalism, the idea that the natural world is the *only* world, and that we can use the tools of rational inquiry to understand it.

Today, the case for naturalism is rooted in one of the most mysterious and counterintuitive areas in all of science: the strange world of quantum mechanics, which is where we'll journey next.

CHAPTER 3

Quantum Theory and the Power of Naturalism

If you think you understand quantum mechanics,
you don't understand quantum mechanics.

– Richard Feynman

I n 1801, Thomas Young, a British physician, conducted a series
of experiments to better understand how the eye perceives
light. Young was a brilliant polymath, described by his
biographer, Andrew Robinson, as "the last man who knew
everything." He was the first person to describe astigmatism; he
discovered how retinas detect color a century and a half before its
scientific confirmation; and he provided key insights into deciphering
the Rosetta stone. But perhaps his greatest legacy was the
development of the famous "double-slit" experiment, in which light
is sent through a barrier with two slits and then projected onto a
screen.

According to the particle or "corpuscular" theory of light,
developed by Isaac Newton a century earlier, this experiment ought
to have created dots that clustered on the screen in in two lines, one
associated with each of the slits, with a thinner distribution of dots
surrounding the lines for random strays. Each dot would be
associated with a single light particle.

But this is not what happened. The photons arrived not in two clusters but in an interference pattern, in much the way waves created by two pebbles thrown in a pond might interact. In some regions, light coming through both slits produced more light, but in some areas it produced *darkness*. Thomas had demonstrated that light behaves like a wave.

This, in itself, was astounding. But, as later scientists discovered, the real implications were far more profound. Photons, like all objects, have both a velocity and a position. Curiously, if you observed one aspect of the photon – say, its velocity – it was impossible to also know its position. Moreover, measuring one of these attributes seemed to affect what results you'd get. The implication that emerged was truly revolutionary: the sub-atomic world didn't seem to behave like anything like the mechanistic macro world.

In 1897, physicist J.J. Thompson discovered the electron, which was initially thought of as a little lump of electrically charged matter that zoomed around in predictable ways, placing it squarely in the particle camp. Around the same time, Max Planck in Berlin was studying the characteristics of energy emitted in the form of radiation – such as the heat emerging from a fireplace. Planck discovered a formula that described his experimental results well. But there was something strange hidden in the math. The units of energy that fit into his equation always came in discrete whole numbers, such as 1, 2, or 3. You'd never find an energy of, say, 2.7 or 5.3. Weird.

In 1905, along came a young patent clerk named Albert Einstein. Einstein found that in some ways light seemed to act as a wave, but in other ways as a rain of tiny corpuscles, or particles. Einstein called these little units of light *quanta*.

He relayed his new findings in a famous letter to his friend Conrad Habicht:

So, what are you up to, you frozen whale, you smoked, dried, canned piece of soul, or whatever else I would like to hurl at your head, filled as I am with 70% anger and 30% pity! You have only the latter 30% to thank for my not having sent you a can full of minced onions and garlic after you so cravenly did not show up on Easter. But why have you still not sent me your dissertation? Don't you know that I am one of the 1 1/2 fellows who would read it with interest and pleasure, you wretched man? I promise you four papers in return.

Habicht was on the receiving end of what was surely one of the finest barters in the history of science. The "inconsequential babble" of Einstein's letter refers to four new papers he was working on, including one on "radiation and the energy properties of light," which he described as "very revolutionary."

Over the next several decades, the insights of Young, Thompson, Einstein, and others coalesced into a bold new theory. Particles don't "exist" or "not exist" – they have a *probability* of existing. Particles may appear or disappear into our universe according to some random pattern, but they have no apparent cause. These findings were so troubling to Einstein that they inspired him to issue on of his most famous quotes: "God does not play dice with the Universe."

Quantum mechanics and naturalism

Quantum mechanics is one of the most mysterious and mind-boggling areas in all of science. Its findings upended our understanding of the natural world and challenged many deeply held notions, such as the simple cause-and-effect dynamics of Newtonian mechanics. But in doing so, it reaffirmed and strengthened the case for naturalism – the principle that the natural world is the *whole* world, and that science an incredibly effective way of understanding it.

Quantum field theory provides the technical foundation for contemporary naturalism. It gives us an extraordinarily accurate understanding of how the physical world works at the most elemental level. The other elements of reality are built upon this structure. As matter interacts at the quantum level, it creates more complex features, which we call molecules. Molecules are simply the aggregate of the particles which comprise them, each of which behaves according to defined rules. More complex patterns of behavior – from weather patterns to evolution to the brain states that underly our very emotions – derive from the layers underneath them. With perfect information about the world, we ought, in theory, to be able to describe how a brain works, or how a society works, based on the aggregate behavior of its component subatomic particles.

Naturalism as a worldview leaves some room for interpretation. It is sometimes defined in terms of what *tools* may be used to understand reality – such as experience, reason, and science – as opposed to a description of the true nature of reality. All naturalistic viewpoints, though, assume that science has at least some explanatory power about the world.

The physicist Sean Carroll defines naturalism based on three premises:

1) There is only one world, the natural world.

2) The world evolves according to unbroken patterns, the laws of nature.

3) The only reliable way of learning about the world is by observing it.

This seems to me as good a definition as any – though the first premise may mean something much more profound than Carroll intended. We'll soon see that the "natural" world is the home of not just particles and planets but of spirituality itself.

Philosophical implications of quantum theory

Quantum mechanics is often used to illustrate philosophical principles about the nature of existence, sometimes accurately and sometimes speciously. So it's worth taking a brief side journey to understand what, exactly, is happening in this strange quantum reality, and to examine its implications for our spiritual journey.

One of the earlier attempts to explain quantum mechanics was the Copenhagen interpretation, devised between 1925 and 1927 by Niels Bohr and Werner Heisenberg. It says that objects remain in a permanent wave state until they are observed. Physical systems do not have definite properties prior to being measured, and quantum mechanics cannot determine with certainty what will happen, but can only predict the probabilities of observing certain results. The act of measurement itself affects the system, causing the set of probabilities to reduce to only one of the possible values. When measured, the wave function collapses into a single reality.

The Austrian physicist Erwin Schrödinger illustrated this concept with a famous thought experiment. Imagine a cat trapped in a steel box which contains a vial of radioactive substance and a vial of poison. If a particle of the substance decays over the course of an hour, it triggers a hammer that shatters the vial, thereby poisoning the cat. According to the Copenhagen interpretation, if we do not measure the system at any point in the hour, the particle is in a state of quantum superposition, and thus exists as a combination of states at the same time. And that means that the cat itself also exists in a combination of states: it is *both* dead and alive until we open the box, at which time we observe the situation and the cat's wave function collapses.

The Copenhagen Interpretation is no longer widely accepted by scientists. Einstein complained to Bohr: "Do you really think the moon isn't there if you aren't looking at it?" (Bohr, in response, said "Einstein, don't tell God what to do.") The currently prevailing idea is that the indeterminate wave function actually reflects our own lack of full knowledge of the system, not the fact that the particle is really in two states at once. In the case of Schrodinger's cat, it's not true that the cat is 50% alive and 50% dead until we observe it. It's simply that there is a 50% chance it is dead and 50% chance it's alive.

There have been some attempts to reconcile these two worlds – the indeterminacy captured by the Copenhagen interpretation and Einstein's intuition that a physical state must either exist or not exist, with no wiggle room in between. The *Many Worlds Interpretation*, or MWI, was popularized by Bryce Seligman Dewitt in the 1960s and 1970s. According to the MWI, when a wave function collapses, it's actually creating two separate worlds. Thus, every possible world is actually real, and we just happen to live in one of them. All existence is made up of an unfathomably large set of universes, each in quantum superposition. These universes would be unable to communicate with each other, and thus impossible to detect.

A related, biocentric idea is the *Many Minds Interpretation* (MMI). Each time an observer interacts with a quantum process and the wave function collapses, rather than pinching off two different universes, two different awarenesses are created. According to the MMI, it's not the worlds that branch, it is the observers' minds, which actually select a specific quantum state. As a result, there are an infinitely large number of Andrews out there, one for each possible universe.

There is a clear problem with this idea, beyond the obvious horror of an infinite number of Andrews. The MMI relies on an observer to make the quantum mechanics go, but what happened before there was conscious life? Do wave functions collapse in the absence of conscious observers?

Randomness and order

You might wonder whether the randomness of the quantum world upends my declaration that the world is fundamentally ordered. Quantum processes are usually assumed to be truly random, but *random* does not mean the same as *disordered*. The random quantum breakdown of the isotope oxygen-19, for example, actually exposes an underlying pattern. Its half-life – the amount of time it takes for half of the atoms of oxygen-19 to decay – is precisely 26.464 seconds. Take any quantity of oxygen-19 anywhere in the Universe, and half of its atoms will decay in 26.464 seconds. One atom may act completely randomly, but together, they seem to act according to a shared purpose, as if they are secretly communicating with each other.

While the behavior of individual atoms is random, the behavior of groups of atoms is highly predictable. If I drop a Ming vase on a concrete floor, it's possible that any given atom might simply decay into nothingness. But when summed across all the atoms in the vase, it's a virtual certainty that the force of gravity will send the vase hurtling to Earth.

Systems that appear random actually reflect a deterministic order. Consider, for example, a chaotic system, like the world's climate. The system appears random because tiny variations can have large-scale impacts – the so-called "butterfly effect." Even a seemingly mechanistic billiards table appears to behave chaotically. The physicist Michael Berry calculated that the movement of a single electron on the other side of the Milky Way is enough to cause a tiny fluctuation in the movement of a billiards ball on Earth – enough to fundamentally change the outcome of the game.

Similarly, the behavior of a roulette wheel appears random, but the system's complexity belies an underlying predictability. Building on this fact, a group of scientists in the 1980s were able to predict with some accuracy where the ball eventually landed by measuring factors like the velocity of the ball and the rotation of the wheel.

The upshot is that only at the tiniest quantum level is our world random. At the level of human experience, all is determined according to the laws of nature of which we are a part.

Quantum mechanics and the "physical" world

Quantum theory reveals another important observation about how our brain perceives and processes reality. For most of human history, the reality of the physical world seemed unassailable. I could touch an object in the physical world – say, a rock – and it seemed to have its own, independent essence, apart from what my mind created it to be. Quantum mechanics demonstrates that the world we perceive is not the "real" world; it is a *story* about the real world. This story includes concepts like "matter" and "causation." Quantum mechanics shows us that the "real" world is quite different. In fact, even our understandings about the quantum world are stories, using metaphors like "waves" and "particles."

This principle extends to all of our assumptions about the natural world. The concept of "color," for example, exists nowhere in the physical world. What we call color is a story created by our brains when light waves of various lengths hit our retinas and our optic nerves convey that data to our brains, which in turn creates an experience that we call "color." Even this story has a story within it: a "light wave." This idea is simply a metaphor we've constructed to help us get a handle on something fundamentally ineffable.

We can't directly perceive, through rational thought, the raw truth of nature. Our brains are like CD players; they take individual bits of information and convert them into a story that makes sense to us. Trying to understand what the quantum world is "really" like is like examining the surface of a CD and expecting to experience music.

The stories our brain creates are not exactly "illusions," and they are not "wrong." They are necessary constructions to allow us to make sense of the raw data of nature. But even *this* observation is in itself a story, for the terms "data" and "nature" are themselves stories we have created. Reality just *is*.

Implications of naturalism

The opposite of naturalism is usually taken as supernaturalism, so naturalism is typically associated with atheism. Daniel Dennett, one of the philosophers mostly closely associated with atheism, disavows anything supernatural, from the Easter Bunny to God to life after death, calling himself and others of like mind "brights." "'I'm a bright' is not a boast but a proud avowal of an inquisitive world view," he writes. Brights are the "moral background of the nation."

A criticism of this position from religious quarters is that divine action, rather than the laws of nature, can help explain certain phenomena. For example, the theologian Alvin Plantinga suggests that quantum mechanics can help explain Jesus turning water into wine. This is simply not a good description of how the world works, explains Carroll. This seemingly miraculous occurrence would be *allowed* under quantum mechanics, but is so extraordinarily unlikely that in repeated universe-size experiments a single drop of water would most likely not be turned into wine.

Carroll's point is that the laws of physics are very well tested and perfectly describe our everyday life. Divine intervention or paranormal phenomena *could* exist, but would have to violate laws of physics which have been experimentally confirmed. Science compels us to pick the more likely explanation for an occurrence. Since we don't see experimental evidence of divine action or paranormal phenomena, and since they violate the laws of physics, there is no reason to postulate it – except that we would like it to be true.

Naturalism is closely related to the controversial belief known as "physicalism." This school of thought holds that everything – mind, spirit, body, rocks, and trees – are fundamentally made out of the same stuff, which is physical. Physicalists believe that the workings of the physical world describe all that is, from the ocean's tides to our own passions and desires. It holds that every phenomenon is caused by the interaction of matter across space and time. That's it, nothing more.

To some physicalists, even consciousness itself is simply a physical phenomenon. T.H. Huxley, the nineteenth-century biologist and ardent Darwinian, called the stream of consciousness an "epiphenomenon" of evolution. Like the bell of a clock that has no role in keeping time, our consciousness does not actually determine behavior. "The consciousness of brutes," he wrote, "is simply a collateral product of its working ... as the steam-whistle which accompanies the work of a locomotive engine is without influence upon its machinery ... The soul stands to the body as the bell of a clock to the works, and consciousness answers to the sound which the bell rings out when it is struck ... We are conscious automata."

To epiphenomenalists, consciousness, like the steam-whistle, reflects what's going on inside the engine, but does not actually cause the train to move. Our brain states, and therefore our desires and emotions, are simply manifestations of matter. The neuroscientist V.S. Ramachandran writes, "all the richness of our mental life – all our feelings, our emotions, our thoughts, out ambitions, our love lives, our religious sentiments, and even what each of us regards as his or her own intimate private self – is simply the activity of these little specks of jelly in our heads, in our brains. There is nothing else." Or in the pithy words of the philosopher Daniel Dennett, "In short, the mind is the brain."

The profound implication of this position is that even our emotions and experiences – our *qualia* – are fundamentally physical. Our brain stores emotions and memories in its complex set of neurons and in the connections between them. Even memories are not really memories, in the familiar sense of a direct window to a previous reality. They are physically instantiated patterns in our neurons which may or may not reflect events that actually happened. Memories are not ineffable, immutable parts of "us" that are somehow attached to our non-physical spirit, but rather physical structures in our brains which are mutable, changing as the very molecules in our brains change.

So everything is physical, and that's all there is? For all its rational justification, naturalism, you'd be reasonable to object, is a depressingly reductionist view of the world. If everything is reducible to the mechanistic behavior of subatomic particles, do such human concepts as *love* or *morality* or *God* have any meaning at all? Are our brains simply – to put it in the brusque phrasing of computer science pioneer Marvin Minsky – meat machines?

Yes, naturalism is a powerful account of reality, but it's only *part* of the story. To get to the rest, I'll need to get away from the textbooks and back on to the trail.

CHAPTER 4

Mystical Naturalism

The eye through which I see God is the same eye
through which God sees me; my eye and God's eye
are one eye, one seeing, one knowing, one love.

— Meister Eckhart, Sermons of Meister Eckhart

I t's an apple-crisp fall afternoon and I'm gliding down the Wapiti
Trail, a serpentine swatch of singletrack in the foothills below
Longs Peak. I've been training all spring for a 100-kilometer
ultramarathon in the Never Summer wilderness, just a few hours from
here. All those hours climbing up and down these mountain trails
have finally started to pay off, and my body feels strong and free.

I'm polishing off the final segment of an hour loop, descending
through a glade of ponderosa pines to an open meadow. Across the
valley rises the majestic summit of Bear Peak, its flanks of evergreen
trees turned a warm sage in the afternoon sunlight.

As I gain speed on this last descent, I hook into that elusive
feeling of here-ness. I'm part of everything – the trees and mountains
around me, the trail, my body, my mind, the very laws of nature. For
this moment, we are all one together. It's one of those perfect
moments that sometimes come along on a run, a little gift of true
connection and presence. For a moment at least, the analytic scientist
in me turns mystic.

The mystical experience is part of virtually every cultural and religious tradition we know of. Followers of eleatics, a school of philosophy founded by Parmenides in the sixth century BCE, believed in the essential unity of all things. The classical Greeks called this concept *henosis*, translated as mystical "oneness," "union," or "unity" with the One. A babbling brook was an expression of the limitless energy of being or existence.

Plato believed the world itself was a living thing with a soul and intelligence. Later, the neoplatonists and Christian mystics saw henosis an ecstatic experience of joining with God. We all emanate from and return to the One, the source and end of all things. All living things are connected by the *anima mundi,* or the "world soul," an intrinsic connection between all living things on the planet.

The anima mundi is similar to the Jewish concept of *Chokhmah Ila'ah*, a supernatural wisdom that transcends, orders, and vitalizes creation. It finds a cousin in the early Christian concept of *mystikos,* which referred to three intertwined dimensions: the biblical, the liturgical, and the spiritual or contemplative. The fourteenth-century work of Christian mysticism evocatively titled *The Cloud of Unknowing* describes how spiritual union with God can be achieved by emptying the mind of images and thoughts.

In Eastern philosophy, mysticism is closely associated with the idea of non-duality and death of the self. In various forms of Buddhism, all things are seen as inter-related and mutually dependent, an idea expressed by the concept of s*unyata,* which simply means "emptiness." In the West, we think of emptiness as a void needing filling, which we fill with material objects or with the ongoing babble of our endlessly chattering minds. In Buddhism, emptiness itself is the goal. It refers to a liberation from the clutter that is not truth itself, but impedes us from experiencing truth. It is the understanding that there is no "self" that is not part of the greater whole. When we truly understand *anatta*, the doctrine of not-self, we will have attained the state of *nirvana*, a Sanskrit word which roughly means "disappearance" or "extinction" – the extinction of the illusory belief in a self separate from all reality.

In Hindu thought, mysticism roughly refers to a union with God (unio mystica), but even the word "union" doesn't quite fit, because it implies *two* things coming together. For example, in the Hindu school of *Advaita Vedanta*, there is only one reality, Brahman. The essence of Brahman lives in all of us in the form of *atman* – literally, "breath." Non-duality is also at the heart of the Indian school of *Madhyamika*, in which an ultimate nondualistic reality exists beyond the apparent duality of conventional and absolute truth; and it is underlies the Zen Buddhist concept of *rigpa*, or nondual awareness. In Tantra – an ancient set of mystical traditions of Hinduism and Buddhism – humans are not meant to transcend the world, but to absorb themselves into it, finding bliss by experiencing the divine within one's own body.

The idea of the nondual soul is artfully expressed in the parable of Shvetaketu in the ancient Hindu texts the Upanishads. The Vedic sage Uddalaka, in his ongoing quest to try to convey the truth of existence to his arrogant son Shvetaketu, tells him to throw salt into a bowl of water and return in the morning. The next day Uddalaka asks Shvetaketu to retrieve the salt, but of course his son cannot find it, as it has dissolved. And yet the water tastes salty. "You cannot make out what exists in it, yet it is there," says the teacher. "Likewise, though you cannot hear or perceive or know the subtle essence, it is here. Everything that exists has its Self in that subtle essence. It is Truth. It is the Self, and you, Shvetaketu, *you are That.*" Our souls are salt in the great cosmic water bowl.

The mystical experience

Mystical experiences almost always have one thing in common: a state of union or oneness. "In mystic states we both become one with the Absolute and we become aware of our oneness," said William James. Rudolf Otto, the early-twentieth-century theologian, set as his life's work to explore the "numinous," the profound emotional experience of connection with the spiritual realm. He called mystical experience the *mysterium tremendum*, describing it vividly: "It may at times come sweeping like a gentle tide, pervading the mind with a tranquil mood of deepest worship. It may pass over into a more set and lasting attitude of the soul, continuing, as it were, thrilling vibrant and resonant ... it may burst in sudden eruption up from the depths of the soul with spasms and convulsions, or lead to the strangest excitements, to intoxicated frenzy, to transport, and to ecstasy."

The mystical has entranced modern thinkers like the sci-fi author Rudy Rucker, who observed, "The central teaching of mysticism is this: *Reality is One.* The practice of mysticism consists in findings ways to experience this unity directly. The One has variously been called the Good, God, the Cosmos, the Mind, the Void, or (perhaps most neutrally) the absolute. No door in the labyrinthine castle of science opens directly onto the Absolute. But if one understands the maze well enough, it is possible to jump out of the system and experience the Absolute for oneself."

The mystical experience is an inherent part of the human experience. But where does it come from? Why are we endowed with this mysterious bridge between us and the divine? What purpose does it have, if any? Many people who've had mystical experiences have a sense they have been let in on a deep secret of the Universe, what William James calls "the noetic quality." Is this secret "real?" How would we know?

One possible answer is that mysticism is a direct proof of the existence of God. Mystical experiences, the argument goes, show us with indisputable surety the existence of a supreme being so undeniably real that no further proof is needed.

Conversely, perhaps mysticism is a side-effect of the natural process of evolution. Maybe, suggests the journalist John Horgan, mystical capacity is a spandrel. In architecture, a spandrel refers to the space between an arch and its surrounding structure. In a 1979 paper called "The Spandrels of San Marco and the Panglossian Paradigm," the Harvard biologists Richard Lewontin and Stephen Jay Gould suggest that human traits are simply functional adaptations to help preserve our genes, and have no purpose in and of themselves. Mysticism is merely along for the ride, a byproduct of genetic adaptation and selection. Awe itself is an adaptation that let the human species persevere by creating positive emotions around the feeling of something being bigger or more meaningful than oneself.

Now, it seems, we are forced to choose: either mysticism is a "real" connection with the divine, or it is merely a product of the natural world.

And here we have it. The whole problem, the entire supposed conflict that separates the two huge masses of belief, the spike that cleaves rationality and spirituality, lies within that one little weaselly word: *merely.*

Merely. As if a process that gives rise to the mystical experience, and by extension to awareness itself, can simply be set aside and forgotten. What if instead of minimizing evolution's spiritual importance, we viewed it as a dialect of the divine tongue of the Universe, expressed in its language of order and mathematics?

Naturalism and supernaturalism

And now we're ready to confront that nasty assumption about naturalism: that if everything is part of the natural world, including our very soul itself, we're left with a spiritual desert, a reductionist emptiness.

To see why that assumption is so wrong, let's start by defining some terms. When we try to precisely define the "natural world" that supposedly comprises all of reality, we see that the term itself has no real independent meaning, but is a "story" we create. This story is comprised of other, component stories. It incorporates ideas like "physical" objects and "laws of nature" – concepts with no objective scientific reality, but rather constructs our brains evolved to help us navigate the world.

What we really mean when we talk about the "natural world" is the whole of ordered reality we can observe through observation and reason. We assume that naturalism must rule out the supernatural – that is, everything that is not ordered and rational. The only alternative to the ordered world is the disordered world. But what in the world could this possibly mean? Such a world has no real meaning. It is identical to chaos. If the natural world is everything that makes sense, then everything that makes sense must be part of the natural world.

One might object that supernatural entities, such as God, may be rational when they want, and capricious when they want. But to *choose* to be irrational is a logical contradiction. To choose to do something, whether it be eating a Skittle or creating a universe, relies on an ordered pattern of cause and effect.

A different, and better, way to unite the natural and supernatural worlds is through the idea of *magic*. Magic, as I use it, does not mean the ability to cast spells or to turn water into wine. It's a qualitative, subjective term encompassing the ineffability of being alive. It is the miracle of existence, the wonder of self-awareness. It is the mystical world.

We've arrived at a radical notion: the natural world is not only compatible with the spiritual world; the natural world *is* the spiritual world. The mystical and naturalist accounts of the world are *both* right, both part of the same unified vision of the world.

This is a vision I call *mystical naturalism*.

It holds a vision of naturalism that is not opposed to supernaturalism, but which subsumes the supernatural within itself. Naturalism is not *reductionist*, in that it reduces everything to natural explanations, but *expansionist* – in that it expands our view of the natural world into the realm of spirit.

Mystical naturalism neatly dismantles the most devastating critique of science: What if science, in its infinite wisdom, sucks the very magic out of the world, blinding us to the miracle of existence? What if, once science is done, all we have is a bleak, purposeless nothingness?

To diffuse this complaint, it's worth taking a close look at that single word: "miracle." The dictionary definition of a miracle is "an extraordinary and welcome event that is not explicable by natural or scientific laws and is therefore attributed to a divine agency." A miracle is something that, by the conventional definition, cannot be understood by reason.

But what if we turn our thinking about miracles on its head? Perhaps the world we live in, and the laws that give it order and sense, are themselves wondrous beyond our imagining. In the ever-astute words of Albert Einstein, "The eternal mystery of the world is its comprehensibility ... The fact that it is comprehensible is a miracle." Perhaps the rules of the mystical universe are set just so, to enable the process of evolution to create beings capable of mystical experience.

The idea that the natural world is a miracle is not a fact that can be proven or disproven. It is a state of mind, an orientation. It's a way of seeing and being in the world. We experience it not through logical argument but through wonder, awe, and amazement.

Nondualism

At the bottom of the Wapiti trail, I cruise to a stop on the wooden bridge that passes over a giggling creek. The meadow I've just crossed is shaded cumin and turmeric in the fading sunlight. The immediacy of the run fades as it turns into a set of facts and figures, statistics about miles run and feet climbed. Now its something to write about, to measure, to analyze.

Why is it so hard to get to mysticism, and why does it fade so easily? If it's such an essential part of reality, why do we have such a hard time living in it?

There is a demon at work, one that our brains make possible, but one fortified by our Western culture. This is the demon of duality.

In the annals of philosophy, dualism comes in many flavors. In Western thought, it's most commonly defined as a belief that the physical world and the world of mind are two separate things. Descartes, perhaps the most famous dualist, argued that while our bodies are physical and adhere to the laws of the material world, our minds are not physical at all and are exempt from those laws. Matter is determined, but the spirit of the mind is free. He even suggested a mechanism by which this might happen: that the mind acts on the body via the pineal gland.

William James imagined a physical realm in which our mortal bodies live, and an immortal realm for our consciousness. When we die, "finally a brain stops acting altogether, or decays, that special stream of consciousness which it subserved will vanish entirely from this natural world. But the sphere of being that supplied the consciousness would still be intact; and in that more real world with which, even whilst here, it was continuous, the consciousness might, in ways unknown to us, continue still." Kant, too, imagined two worlds: the world of phenomena and the world of mind.

In the lingo of philosophy, the opposite of dualism is *monism*, the belief that consciousness and matter are made of the same basic thing. Among the early monists were the ancient Greek idealists who believed there was just one thing and that thing was consciousness, or *nous*. They were the first in a long line of idealists who believe that the world is actually made out of mind or spirit, and matter is merely an illusion.

In the twentieth and twenty-first centuries, some philosophers have adopted a more balanced view that admits monism from a Western scientific or religious standpoint. The early-twentieth-century philosopher Alfred North Whitehead updated the mediaeval concept of dualism, in which spirit is eternal and matter is transitory and imperfect, with a more holistic concept of reality. More recently, philosopher John Searle's biological naturalism holds that our entire understanding of physics shows that mind must be made of the same kind of substance as the body.

The various types of monists can fill up a textbook of their own. There are priority monists, who believe all things are derived from The One. There are existence monists, who believe there exists only a single thing, which we call the Universe. There are idealists, monists who believe that the single thing is consciousness, or *nous*; and their opposite, the eliminative materialists, who believe that everything is physical and mental things do not exist at all. And then there are neutral monists, who believe that there is one thing, and it's not quite mind and not quite matter – primary stuff that William James called a "booming, buzzing confusion" of "pure experience." There is *substance dualism*, which says there are both physical and nonphysical substances, and *property dualism*, the idea that physical things have different sorts of properties, some of which can be mental.

All of these distinctions can get esoteric, and on our quest for ultimate spiritual truth we run the risk of getting tangled up in academic arguments and definitions. I'm reminded of what Pirsig called *philosophology*, an academic discipline which "is to philosophy as musicology is to music, or as art history and art appreciation are to art, or as literary criticism is to creative writing. It's a derivative, secondary field, a sometimes parasitic growth that likes to think it controls its host by analyzing and intellectualizing its host's behavior."

I don't share Pirsig's contempt for academic philosophy in general (he never saw the value in being part of a community of thinkers, preferring his reclusive life-of-the-mind). But the distinction between, say, priority monists and existence monists is a semantic one. Getting too tied up in definitions risks unwittingly adopting an existing and potentially limiting frame. Language can help clarify ideas, but sometimes using philosophical jargon actually adds to the confusion, such as when terms mean the *opposite* of what we expect them to mean. The idealist – technically a nondualist – sees no spiritual role for the physical world. This is just the opposite of the dualistic demon I want to exorcise.

Mystical naturalism. It has a sort of ring to it. I see that the natural world is largely understandable through reason and empirical observation and provides the order that makes reality go; thus I am a naturalist. But I am also here and alive; an essential aspect of that reality. Thus I am too a mystic.

At the nexus of mysticism and naturalism we see one of the greatest mysteries in all of philosophy: the enigma of consciousness.

It is consciousness that makes possible the mystical experience, that births the ability to even ponder questions about nature and awareness. And it's there my journey will lead next.

CHAPTER 5

The Mystery of Consciousness

Consciousness cannot be accounted for in physical
terms. For consciousness is absolutely fundamental.
It cannot be accounted for in terms of anything else.

— Erwin Schrödinger

The systematic study of consciousness goes back at least as far as Descartes and the Cartesian dualism named after him. In the centuries since, the scientific study of consciousness has blossomed. It typically investigates questions like these: How are various states of consciousness connected to elements in the physical world, such as substances we've consumed or our sleep patterns? What is the difference between being awake and asleep? How do we sense and integrate information? How do we recall the past and predict the future? What parts of the brain seem to be related to the experience of consciousness?

These questions, expansive as they are, are relatively straightforward. We can use scientific tools, like neuroimaging technology or statistical tests, to come to understand them better. That's why the philosopher David Chalmers calls them the "easy" problems of consciousness.

The "hard" problem, by contrast, is about the very experience of consciousness: our first-person, lived experience of the world, or what philosophers call *qualia*. What is this light inside we call awareness, and why does it exist? Is it just an accidental byproduct of the physical world, or something essential to the very fabric of the Universe —

maybe the *only* thing in the Universe? Is there a universal consciousness, and if so, why do *I* experience it as a separate individual?

My goal on this part of the journey is to understand what science has to say about the nature of consciousness, and to see how far it will take us in understanding the hard problem of consciousness. First off, let's dispense with one rather absurd claim: that consciousness does not exist at all. This is the position of the eliminative materialists, so called because they eliminate all possible descriptions of the world except one: that the world is strictly made out of material stuff. Everything else, including consciousness, is simply an illusion.

Daniel Dennett is the foremost modern proponent of this idea. He observes that consciousness does not reside in one single central place in the brain. He calls this mythical location the "Cartesian Theater" after Descartes, who argued that consciousness requires a soul that exists outside of the physical world. To Dennett, my Cartesian Theater amounts to a theater in my brain where a tiny Andrew (which he calls a homunculus) views the ongoing movie. But inside that homunculus's head must be another theater, and another homunculus, and so on in an infinite and absurd regression. So, Dennett concludes, all consciousness must simply be an illusion.

The problem with this, to put it flatly, is that consciousness is *not* an illusion. An illusion, after all, must be *experienced* by someone – and we call the ability to experience "consciousness." Dennett might say, for example, that the experience of pain is merely an illusion created by our brains. But to *feel* pain is to *be* in pain. The philosopher Galen Strawson calls this denial of subjective experience "The Great Silliness."

Consciousness, we can agree, clearly exists. But how does it interact with the material world?

As we saw in Chapter 4, dualists believe that consciousness is somehow unlinked from the actions of our bodies. In a moment we'll explore some scientific research that illustrates that this is not the case. But we can also use a philosophical thought experiment to see why dualism cannot be right.

Imagine another universe identical to ours in every way – let's call it "Bizarro Universe." It contains the same galaxies and the same planets as our own, and its "Earth" looks identical to ours. The difference is that Bizarro Universe is not endowed with consciousness. The humans on Bizarro Earth have the same biological machinery we have. They eat, sleep, and tell jokes just like us. But they lack the "spark" of self-awareness. (These types of creatures are sometimes called "philosophical zombies").

If this Bizarro Universe could exist, it would mean that consciousness can be unlinked from the physical world, and thus that dualism is true. Alas, this unlinking is impossible. There is no way that Bizarro Earth could exactly resemble Real Earth in every way except that it lacks consciousness.

Imagine I'm a spacetime-travelling philosophical detective and have just landed on Bizarro Earth. I find Bizarro Andrew in the Bizarro universe, writing *Bizarro Science for Seekers*. He's up to this chapter: the chapter on consciousness. I look at the physical pixels on Bizarro Andrew's screen. Will they be the same pixels that are on Real Andrew's screen?

They can't be, because the very idea of consciousness has no meaning on Bizarro Earth, and thus this chapter could not exist in its current form. And just as this chapter of *Bizarro Science for Seekers* could not exist, any discussion of consciousness could not exist on Bizarro Earth. This would have observable, macro effects on the Bizarro Universe. There would be no science of mind, no copies of Descartes's *Discourse on the Method*, and no books on mindfulness, for without a mind there can be no mindfulness.

We know, then, that consciousness has two essential properties: 1) it exists; and 2) it is intimately tied to the physical world. But what *is* it, exactly? Is it some sort of substance, like matter? Is it everywhere, or does it just appear at certain locations and certain times? Is there just one universal consciousness, or is each of our awarenesses eternally and unchangeably separate?

A bold approach to these hard problems is to assert that consciousness is a fundamental element of reality, inherent in the Universe's very existence. This is the doctrine of *panpsychism,* that consciousness pervades reality. Panpsychism is a monistic view of reality because it says that the cosmos is ultimately made of a single kind of stuff. The physicist Max Tegmark even theorizes that consciousness is its own state of matter, like solids, liquids, and gases, which he calls *perceptronium.* This substance can be explained, he says, using five mathematically sound principles.

The idea that consciousness is essential to the Universe, airy-fairy as it might seem, is held by some pretty hard-headed critical thinkers, including the physicists John Von Neumann, John Archibald Wheeler, and Eugene Wigner. Their eponymous interpretation is that thought and consciousness actually *create* matter. According to Wigner, the laws of quantum physics could not be formulated in a fully consistent way without reference to consciousness. For Schrödinger, he of the partially-existing cat, "There is obviously only one alternative, namely the unification of minds ... [I]n truth there is only one mind." Max Planck, the German physicist who was one of the original developers of quantum theory, also regarded consciousness as fundamental to the Universe. Perhaps the Universe, in the words of the English physicist Sir James Jeans, is "more like a great thought than a great machine."

Indeed, there is a sense in which the Universe doesn't exist unless someone is here to observe it. That someone, specifically, is *me*. As long as I can conceive, I cannot conceive of nothing. I can conceive of an empty universe, or a void – but not nothing. The only way I can fathom true nothingness is if I myself can no longer perceive. When I'm gone, in a sense the Universe is gone too.

A variation on panpsychism is offered by the physicist Freeman Dyson, who believed that "mind, as manifested by the capacity to make choices, is to some extent inherent in every electron." The philosopher Thomas Nagel proposed that our brains are made from physical particles, similar to electrons, protons, and neutrons, but which produce subjective thoughts and feelings. Consciousness must somehow lie in those individual particles – so every particle must contain a little bit of consciousness.

The problem with this idea is sometimes called the "combination problem." How many little bits of mind-stuff combine to form a bigger mind? If little bits of matter have a little bit of consciousness, then why don't larger bits of matter – trees or bowling balls or kidneys – have a little bit more consciousness? And why does it seem that one brain is associated with one mind?

The truth is that we just don't know the answers to these questions, and maybe we can never know. Von Neumann et al.'s proposition that consciousness *creates* the Universe is intriguing, but there's no way we can evaluate it in a meaningful way. What we do know is that consciousness somehow embedded into the Universe, and that itself gives us plenty to ponder.

Quantum consciousness

Another approach to explaining the mystery of consciousness is to levy the power and mystery of quantum physics. Talking about consciousness and quantum mechanics in the same sentence inevitably brings the wrath of skeptics. Max Tegmark, a physicist who believes that consciousness could be a quantum state of matter, says, "Consciousness has always been a tricky topic to broach scientifically. In most serious scientific circles, merely mentioning consciousness might result in the rescinding of your credentials and immediate exile to the land of quacks and occultists."

The idea that consciousness might somehow be related to quantum mechanics was first expressed as early as the mid-twentieth century, when the neurophysiologist John Eccles explored the possibility in his 1953 book *The Neurophysiological Basis of Mind.* Eccles examined whether there are locations in the brain where processes associated with consciousness occur on so tiny a level that quantum effects may play a part. He proposed that the synapses, where the axon of one neuron communicates with the dendrite of another neuron, may be just such a place.

The aforementioned Von Neumann-Wigner interpretation is one of the most serious attempts to connect consciousness and quantum mechanics. The idea is that our consciousness itself is what causes collapse of the wave function, converting multiple possibilities into a single reality at the quantum level.

This interpretation, alas, is not accepted by most physicists: a 2011 poll at a quantum mechanics conference showed that only two of thirty-three experts believed in a wave-function collapse caused by consciousness. It would raise several intractable problems, foremost among them how in the world the Universe conducted its business before consciousness came along. If consciousness is necessary for anything to "actually happen," for the world to convert statistical probabilities to reality, how did we get here in the first place?

Perhaps the most acclaimed contemporary proponent of quantum consciousness is the physicist Roger Penrose. In his 1989 book *The Emperor's New Mind*, Penrose expands Eccles's idea that quantum-sensitive neurons in the brain play a role in consciousness. He speculates that many different arrangements of atoms are being "tried" simultaneously, in much the manner of a quantum computer. When tens of thousands of neurons operate together in this manner, perhaps there is an opening for free will to influence these quantum choices. (Penrose's idea is predicated on the false notion that free will requires random behavior. We'll explore this further in Chapter 6.)

Quantum physics has even been martialed to solve the combination problem. Quantum entanglement means that two different particles lose their individual identities and act as a unified system, so perhaps, the thought goes, a brain's worth of entangled particles produces consciousness. Most likely, though, quantum states are simply too tiny and short-lived to affect something on the larger scale of neural processing.

Quantum mechanics and Eastern philosophy

Many modern physicists have seen parallels between quantum mechanics and Eastern philosophy. Fritjof Capra, in his 1975 book *The Tao of Physics: An Exploration of the Parallels Between Modern Physics and Eastern Mysticism*, explored the connections between Eastern thought, particularly Zen Buddhism, and the seemingly paradoxical conclusions of quantum mechanics. The book was based in part on conversations with the physicists Werner Heisenberg and Neils Bohr, who'd both made connections between the mysteries of quantum mechanics and their own mystical experiences.

While Capra did base some of his ideas on conversations with highly regarded scientists, he was criticized in the academic community. Leon Lederman, a Nobel-prize winning physicist and author of the 1993 book The God Particle, said, "Starting with reasonable descriptions of quantum physics, he constructs elaborate extensions, totally bereft of the understanding of how carefully experiment and theory are woven together and how much blood, sweat, and tears go into each painful advance."

Capra and similar thinkers, such as Gary Zukav, who wrote 1979's *The Dancing Wu-Li Masters,* at least did some due diligence in attempting to thoughtfully explore the connections between physics and spirituality, even if their conclusions went beyond what the science of physics could actually support. It was the New Age guru Deepak Chopra who abandoned the scientific foundation underneath quantum mechanics altogether, and used it to "prove" any number of pseudoscientific assertions. Chopra has claimed that his "quantum healing" uses the exact same effects as quantum mechanics to cure aging, cancer, and any number of other maladies. Human aging is "fluid and changeable; it can speed up, slow down, stop for a time, and even reverse itself," as determined by one's state of mind. For his efforts, he was awarded the parody Ig Nobel prize in physics in 1998 for "his unique interpretation of quantum physics as it applies to life, liberty, and the pursuit of economic happiness." (In 2017's *You Are*

The Universe, Chopra takes a more thoughtful approach to such topics, though his ideas generally remain well out of the scientific mainstream).

Unfortunately, pseudoscientific attempts to connect quantum theory and consciousness continue to gain traction, landing in a theater near you with 2004's *What the Bleep Do We Know*, which grossed over $10 million. Produced by J.Z. Knight, a channeler who said that her teachings were based on a discourse with a 35,000-year-old disembodied entity named Ramtha, the movie took the misappropriation of quantum concepts to a breathtaking new high (or low). The theoretical physicist Lisa Randall refers to the film as "the bane of scientists."

The basic motivation behind all this "quantum flapdoodle," to use Murray Gell-Mann's turn of phrase, goes like this: quantum theory is weird and unknown, and spirituality is weird and unknown, so they must have something to do with each other. "No theory in the history of science has been more misused and abused by cranks and charlatans—and misunderstood by people struggling in good faith with difficult ideas—than quantum mechanics," writes physicist and popular science writer Sean Carroll. Victor Stenger characterized quantum consciousness as a myth with no scientific basis that should take its place along with gods, unicorns, and dragons.

Part of the trouble is a long-held misconception of the role of the observer in quantum mechanics. The quantum flapdoodlers – and, to be fair, most of the rest of us – have seized upon the idea that the observer is essential to the process of quantum mechanics. Because the quantum wave function only collapses when observed, we assume that human awareness itself is an essential part of the physical process.

Unfortunately, this is a misunderstanding of the term "observer." A conscious observer is not necessary to collapse the probabilistic wave function into an actual state. All that's needed is something physical to intervene in the process, such as a photon. By way of analogy, imagine a basketball rolling across the floor. You want to determine the location of the basketball by throwing a little rubber ball at it and measuring how long the ball takes to return to your hand.

When your rubber ball bounces off the basketball, it simultaneously returns information about the location of the basketball and *changes* both the velocity and the location of the basketball. The rubber ball is playing the role of the observer, even though it has no awareness itself.

There are more problems with the idea that an intelligent observer is required to make quantum physics go. When, exactly, does the particle's wave function collapse? Is it when the measuring device picks up the signal? When that device's message reaches the retina of a scientist? When the electrical signal is passed to the scientist's visual processing center? What if the observer is a baby? A monkey? The upshot is that the science of quantum mechanics, mysterious and mind-bending as it is, tells us little about the equally baffling phenomenon of consciousness.

Why consciousness?

If the nature of consciousness is mysterious, its purpose is even more so. Why does our universe go to all the bother of generating self-aware beings?

One possible way to answer this question is to explain consciousness as an inevitable outcome of the processes of the natural world. On this view, consciousness is an evolutionary adaptation which confers some benefit upon us, helping us survive and pass on our genes. This is the position of the epiphenomenalists we encountered earlier. For example, psychologists R.F. Baumeister and E.J. Masicampo propose two such adaptations: that thoughts need to be conscious in order to communicate them to others; and that we can only understand complex sequences of ideas if they are conscious. Perhaps we need consciousness to be successful in the social world, such as identifying faces or people in pictures. In a similar vein, psychologist Nick Humphrey argues that consciousness helps us understand how others are thinking so we can cooperate with them to get food and avoid predators.

This position doesn't seem very satisfying. Complexity alone doesn't seem to require consciousness; as far as we can tell the Internet is not conscious, although it rivals the human brain in size and complexity. A good machine-learning algorithm can identify faces in pictures far more quickly than can a conscious human. Epiphenomenalism really just offers an evolutionary explanation of the brain's functioning – one of the "easy" problems of consciousness. It tells us nothing about *why* that ineffable spark of awareness exists, and why philosophical zombies wouldn't do the job of living just as well, without having to write treatises about the mystery of consciousness.

Various thinkers have offered logical rationales for why the Universe must contain consciousness. The British philosopher Derek Parfit suggests that reality is the way it is due to a *selector*, a special feature that causes a particular cosmic possibility to come into existence. A selector is an attempt to explain why this particular reality exists instead of any other. Perhaps, for example, reality follows the principle of Occam's razor. A *razor* is a philosophical principle that allows us to "shave off" alternative explanations for something and come closer to the clean-cut truth. Occam's razor, named after a fourteenth-century Franciscan friar, reads: "More things should not be used than are necessary." In philosophy, it refers to the idea that the simplest explanation for something is usually the best.

Perhaps simplicity is the selector, and the reality we experience is the simplest possible one. Maybe goodness is the selector, and this is the *best* possible universe. Or the selector could simply be fullness – a manifestation of every possible logical possibility.

Or, perhaps, consciousness is the selector. Maybe the Universe "wants" to create consciousness, through the infinite wisdom of an ultimate Mind or through the wisdom inherent in logic itself. What if consciousness simply *must* exist for some deeper cosmic or spiritual reason? This is known as the *teleological* argument, which holds that the Universe works toward some overall purpose. According to the teleologists, the Universe exists for a single purpose: to manifest consciousness. The laws of the Universe were designed specifically to generate self-aware beings, first bringing time and space into existence, then creating particles that joined together in increasing layers of complexity until life emerged, and ultimately, self-awareness.

A variant of the teleological explanation was offered by the philosopher Thomas Nagel his 2012 book *Mind and Cosmos: Why the Materialist Neo-Darwinian Conception of Nature is Almost Certainly False.* Nagel is one of the most well-respected philosophers in the field, but his book created an uproar among materialists.

Nagel's main argument is that consciousness can't simply be explained by the laws of physics alone. Consciousness lies *outside* physics, Nagel believes, so evolution itself – which is responsible for the existence of conscious organisms – must be more than just a physical process. Consciousness must have gotten here somehow, and not through the arbitrary grindings of physical laws. In other words, it has a purpose. The universe "wants" to be conscious. Nagel defends his teleological argument by noting that in addition to physical laws of the familiar kind, there are other laws of nature that are 'biased toward the marvelous.'"

Nagel's argument caused much controversy among the materialist philosopher class. Jerry Coyne, responding to an article in *The Chronicle of Higher Education* defending much of Nagel's thesis, called Nagel's views "pretty much anti-science and not worth highlighting. However, that's The Chronicle's decision: If they want an article on astrology (which is the equivalent of what Nagel is saying), well, fine and good." Steven Pinker tweeted, "What has gotten into Thomas Nagel? Two philosophers expose the shoddy reasoning of a once-great thinker."

I believe there is a problem in Nagel's argument, but it's not that his argument is teleological. Science and reason give us no particular signal as to whether a greater purpose drives consciousness. It's one of those questions that truly is beyond the realm of scientific inquiry.

Where Nagel missteps is in divorcing consciousness from the physical world. Consciousness is not a separate substance in its own right, but is an *aspect* of those laws that make the natural world go. Nagel argues that the laws of physics are not enough to explain the development of consciousness, and so we need more.

According to mystical naturalism, things like spirit and purpose can reside just as happily in the natural world as in the supernatural world Nagel supposes. As our late great friend Robert Pirsig wrote: "The Buddha, the Godhead, resides quite as comfortably in the circuits of a digital computer or the gears of a cycle transmission as he does at the top of the mountain, or in the petals of a flower. To think otherwise is to demean the Buddha - which is to demean oneself."

The mystery of the self

In his 1637, a 41-year-old French philosopher published a treatise that would change the course of Western philosophy. In his "Discourse on the Method of Rightly Conducting One's Reason and of Seeking Truth in the Sciences," René Descartes wrote three Latin words that seemed unassailable: *Cogito ergo sum*. The phrase is usually translated to English as "I think, therefore I am," but Descartes meant something subtly different. He meant not that thinking itself causes existence, but that the fact that I think and I exist are necessarily true together – more like "I think, so of course I am."

The Cartesian view expresses an assumption we walk around with every day: that my unique "I" is an incontrovertible, irreducible phenomenon of reality. Moreover, it assumes there is exactly one "I" associated with each human, and one human associated with each "I." These assertions seem so self-evident as to scarcely require mentioning. After all, every thought must have a thinker, right?

If we dig a bit deeper we see that Descartes's most dearly held assumption starts to crumble. The irreducible "I," it turns out, is another one of those stories our brains have created to make sense of the raw data of the world.

One way to unpack this is through thought experiments, like the one proposed by philosopher Bernard Williams in 1970: what if you were informed you were going to be tortured tomorrow, but prior to the torture, you were told your memories would be wiped out by the neurosurgeon and replaced with my memories. Would you still have reason to fear the torture? In other words, would it still be *you* experiencing the torture?

The answers depends on how we define that tricky word "you" – the self. One way to define the self is by what Williams calls the *psychological* criterion, in which saying "I exist" asserts "the existence of a certain more or less continuous bundle of memories, perceptions, thoughts, and intentions. What makes me *me* and you *you* is our distinctive bundles." The *physical* criterion, on the other hand, says that my identify as a self is determined by my brain.

What the self is, it seems, depends on how we define it.

Derek Parfit's famous Martian experiment gets at the same point. Imagine you step into a machine that reads the location of every atom in your body and transmits the information to a receiver on Mars, which then recreates the body exactly as it appeared on Earth and destroys the original. Would that Martian be "you"? It would have all of your memories and beliefs. It would think it's *you*. But would it really be you? To further complicate things, what if the machine does not destroy the original you? Now there are two "you"s. Which one is the "real" you, and which is the impostor?

The Martian experiment challenges the idea that our own consciousness is somehow associated with a particular combination of materials in the physical world. Even if we are not being teleported to Mars, though, it's easy to see that the self cannot be uniquely associated with some state of the physical world. The cells in our brains grow and die all the time, and the atoms that make up those cells change over time. On the smallest, most basic level, the physical stuff that makes up my brain is completely different than it was a day or a month or a year ago. My atoms are constantly changing. So if my atoms are identical to *me*, then I am slowly changing into a new person.

For example, I am very similar to the Andrew typing on this screen one minute ago, but I am not precisely that person. That person exists at a slightly different point in time, with slightly different understandings of the world. That person's brain and body are slightly different from the Andrew of now. I am *mostly*, but not *exactly*, one-minute-ago Andrew.

Now how about the Andrew of October 22, 1970 – one day after "I" was born. How similar is that Andrew from Current Andrew? Well, the body of Current Andrew is completely different: 98% of the atoms in my body are replaced every year. Some parts of myself, like my skin, gives itself a complete replacement every month.

What about my brain, the location I usually consider the seat of my unique self? The atoms in my brain are arranged in many of the same patterns as the atoms in One Day Old Andrew's brain. But much of that baby's brain will look altogether different than mine. It will not contain any of the memories of my life, my knowledge of language, or any of the zillions of things I've learned over the past five decades.

As the famed physicist Richard Feynman said:

> "So what is this mind of ours: what are these atoms with consciousness? Last week's potatoes! They now can remember what was going on in my mind a year ago — a mind which has long ago been replaced. To note that the thing I call my individuality is only a pattern or dance, that is what it means when one discovers how long it takes for the atoms of my brain to be replaced by other atoms. The atoms come into my brain, dance a dance, and then go out — there are always new atoms, but always doing the same dance, remembering what the dance was yesterday."

If my thoughts really are last week's potatoes, then what do I have in common with One Day Old Andrew, apart from some patterns of neurons that have been passed along through the years, and some consistent external markers, such as whom I identify as my parents, my eye color, and so forth? Perhaps, as Parfit has argued, personal identity doesn't matter at all. At Time 1, there is a person. At a later time, Time 2, there is a person. These people seem to be the same person, because they share memories and personality traits. But based on what we know about the physical world, there is no logical reason to conclude that they are, in fact, the same person. Our brains have created the psychological illusion of continuity of the soul.

In important ways, the Andrew of this moment is a different person than the one of a moment, or a year, earlier. But while all these Andrews are different in some way, they do have some intimate relationship with each other. If Future Andrew is a wholly separate person from Current Andrew, then I have no reason to be more worried (or excited) about my future self than about others' future selves. It would not make sense to withhold pleasures in the moment, such as drugs, in order to benefit Future Andrew.

Beyond logical proofs like these, its hard to get around the intuitive feeling of connection with our past and future selves. So what if, instead of concluding that every person at every different moment is a wholly different, atomized self, we came at a solution in the opposite way? Perhaps we are all part of the *same* self, expressed in billions of different instantiations. We tend to group these instantiations together based on their apparent continuity across time, and call the whole sequence a *soul*. But the real soul is the *anima mundi*, the cosmic consciousness we are all part of.

Some of the most fascinating challenges to the idea of a monolithic self have come from neurological experiments, such as observations of split-brain patients. Patients with epilepsy have sometimes been treated by severing the corpus callosum, the collection of nerve fibers which communicates between the left and right hemisphere. In one experiment, the researchers showed a split-brain patient two pictures, one of a chicken's claw and another of a house in winter. The house was positioned so it would only be visible through the patient's left visual field (corresponding to the right hemisphere), while the claw was visible only to the right visual field (corresponding to the left hemisphere).

Next, the researchers showed the patient a series of pictures and asked him to choose one picture with his left hand and one with his right hand. When the patient saw a picture of a snow shovel and a chicken head, he took the snow shovel with his left hand (which corresponds with the right hemisphere, which has also perceived the wintry house) and the chicken head with his right hand (corresponding to the chicken claw). When asked why he made those choices, the patient explained, "The chicken claw goes with the chicken head, and you need a snow shovel to clean out the chicken shed." The results seem to show that the patient's left and right hemispheres aren't communicating at all. The left hemisphere's job is to interpret data from both visual fields. But it has no knowledge of the snowy house, so it must concoct a logical reason for why the shovel was chosen.

In another experiment, the Finnish neuroscientist Antti Revonsuo followed a similar procedure, showing patients a flower in the right visual field and a rabbit in the left visual field. Since the right visual field, which connects to the left hemisphere, is dominant, the patient reports only that they saw a flower – even though they can point to the rabbit with their left hand. Revonsuo's conclusion was profound. He believed that patients in these experiments are exhibiting *two* consciousnesses, each with their own desires and experiences.

Other insights into the nature of the self come from studies of spiritual practices that affect our brain, especially meditation and the use of psychoactive drugs. This is the realm of neurotheology, which explores the relationship of activity in the brain to spiritual experiences.

Almost every religious tradition contains some type of meditation practice. Meditation may be used for general relaxation, for visualizing or goal setting, and for reducing anxiety and mental stress. But it may also be used to pursue deeper spiritual goals, such as connecting with the deity of your choice or to achieve "enlightenment," often defined as the loss of a sense of self and experience of unity with the rest of the world. To the mystic Rudolph Steiner, meditation is "what unites us with the spirit, the way to knowing and beholding the eternal, indestructible, essential center of our being."

In the 1990s, Andrew Newberg and Eugene D'Aquili, staff at the University of Pennsylvania, scanned the brains of eight American Buddhists practicing a form of Tibetan meditation and three Franciscan nuns engaged in contemplative prayers. The results showed patterns distinct from regular day-to-day brain activity. Most of the Buddhists and nuns displayed increased neural activity in the prefrontal cortex and decreased activity in the posterior superior lobe, demonstrating, in the account of the researchers, that mystical experiences are "neurologically real."

Meditation seems to work by shutting down the part of our brain that is constantly talking to us and about us. This is known as the Default Mode Network (DMN), and it confers upon us some important functions, including self-referential thinking, moral reasoning, and empathy. It is the narrator that talks us through our day, the "editor" that takes in information from other regions and orchestrates a coherent response. When the DMN is turned off, the construct we think of as our "selves" disappears.

Meditation is not the only way we can shut down our brain's DMN. Another is through psychoactive drugs, which are currently enjoying a cultural and scientific resurgence. The modern psychedelic drug movement began in 1938, when a young chemist at a pharmaceutical firm in Switzerland named Albert Hofmann began experimenting with a class of substances derived from ergot, a fungus that infects grain. He called one particular derivative of this chemical lysergic acid diethylamide, or LSD. In spite of his usually rigorous safety procedures, Hofmann somehow managed to absorb some through his skin and experienced the world's first acid trip.

In the early 1960s, psychoactive drugs were closely identified with the counterculture movement, and were the subject of some promising clinical research, most notably by Timothy Leary at

Harvard University. In the early 1970s, the backlash to the counterculture movement, led by Richard Nixon's anti-drug crusade, led to psilocybin being classified as a Schedule 1 drug in the United States, putting it in the same legal category as heroin. Psilocybin, however, has none of the ruinous effects of "hard" drugs. It is not addictive, and "bad trips" are rare under controlled circumstances. In fact, in a recent study of 1,000 patients, not a single adverse event was reported.

As its taboo nature has worn off, psilocybin has enjoyed something of a scientific and therapeutic renaissance. The results of controlled studies on the substance are remarkable. In one study in which psilocybin was administered to patients with terminal cancer, fully two-thirds reported the experience as one of the five most meaningful of their lives, and a third reported it as *the* most meaningful. Many of those who received a high dose of psilocybin had a mystical experience, described as a dissolution of the ego, accompanied by a sense of merging with nature or the Universe.

In *How to Change Your Mind*, a chronicle of the new psychedelic revolution, Michael Pollan writes that taking psilocybin provoked a profound questioning of his own identity:

> So who was this other I? Good question. It wasn't *me*, exactly. Here, the limits of language become a problem: In order to completely make sense of the divide that had opened up in my perspective, I would need a whole new first-person pronoun. And then, having acknowledged the squishy new terrain of identity onto which we had stepped, I went on to characterize this bare disembodied awareness, which gazed upon the scene of the self's dissolution with benign indifference. I was present to reality but as something other than my self. ... There was life after the death of the ego. This was big news.

LSD, psilocybin and other psychedelics, like DMT, are sometimes called entheogens, which refers to a psychoactive substance that induces spiritual experiences in the user. By quieting the part of our brains that insists on creating a story of an independent self, they illuminate the underlying nondual nature of awareness.

Whether these experiences are "factual" or not is a matter of interpretation. Some users of psychedelic drugs have claimed encounters with other realms or beings, or insights into heretofore hidden mysteries of existence. As with supernatural fact assertions, there is no reason to attribute particular weight to these claims. We don't need to, anyhow. The very spiritual experience engendered by entheogens is an existence proof: we can have a mystical experience of oneness, so the world must be mystical.

Consciousness and the limits of science

Thought experiments and scientific observations alike challenge our notion of a separate and indivisible self persevering over time from the instant of birth to the moment of death. In truth, we can probably never understand with certainty what consciousness is and whether we are all part of the same cosmic consciousness, or if there is something meaningful about the individual self. The inescapable conclusion is that consciousness is far less understood than we think. It remains perhaps the most perplexing phenomenon in all of science, even in all of philosophy. "To determine what modes or actions light produceth in our minds the phantasm of colour is not so easie," said Isaac Newton. Steven Pinker says simply, "We have no scientific explanation [for consciousness] ... The brain is a product of evolution, and just as animal brains have their limitations, we have ours. Our brains can't hold a hundred numbers in memory, can't visualize seven-dimensional space."

In 1875, William James drew a deeper philosophical implication from this human inability to truly understand consciousness. Just as a dog will never be able to understand human language, the human mind may be forever closed to certain aspects of the larger universe. For example, the meaning of suffering due to a vivisection is inaccessible to the dog, but that does not mean the vivisection is meaningless. So it may be with our suffering in this world.

The linguist Noam Chomsky distinguishes between *problems* that seem solvable through scientific methods and *mysteries* that do not seem solvable even in principle. He is among the New Mysterians, named after the '60s psychedelic band ? and The Mysterians. This group of philosophers, which also includes Roger Penrose, Steven Pinker, Thomas Nagel, Sam Harris, and David Chalmers, believe that the hard problem of consciousness can never be resolved by humans, even in principle. We just need to accept consciousness as an irreducible element of reality, just as we accept matter and energy.

The power of science is not in telling us everything there is to know about reality, but in telling us what is *not* true about reality. It excels at rejecting false claims and adjudicating between different explanations of reality. But science and critical inquiry have their limits, and we ought to proceed with humility when we reach beyond them.

We're at the boundaries of scientific inquiry, a cliff's edge we can look out across, into the vastness, but may never be able to fathom. But we've made some important progress on our scientific seeker's journey. We've seen that the traditional notion of the indivisible, dualistic self is a shaky one, challenged by scientific research, logical experiments, and direct experience.

And that insight will be essential for the next phase of the journey: unlocking the centuries-old conflict between determinism and free will.

CHAPTER 6

Determinism (vs.?) Free Will

A man may be a pessimistic determinist before lunch
and an optimistic believer in the will's freedom after
it.

— Aldous Huxley

I'm standing at the edge of a dirt parking lot in the middle of a
craggy stand of trees and studying the trail map for Mt. Belford,
which I plan to climb this morning. Mt. Belford is one of
Colorado's forty-eight 14'ers, rising over 14,000 feet into the pale
summer sky above the Collegiate Range. It's a significantly higher
ascent than my run to Bear Peak's summit, and it's steeper, too,
climbing 4,500 feet in under six miles.

I adjust my running pack, which is filled with a bladder of water,
some energy gels, and a few lightweight treats for the trail, and start
up a series of long switchbacks winding through a thatch of aspens. I
settle in for a long morning ahead, and the deliberate, steady rhythm
of climbing allows my mind to wander. I often feel a sense of
philosophical freedom on long mountain adventures like these, with
plenty of time to let connections arise as they might.

Today I'm mulling over a question that's been gnawing at me for
a while: Do humans have free will, or is our behavior determined by
the physical world? On the surface, it seems a standard-issue
Philosophy 101 topic to argue over a couple of craft beers in the
campus pub. But as I examine this conundrum, turning it this way and

that, examining it for structural weaknesses and metaphysical clues, it's clear the question is far deeper and more significant than I'd ever realized.

A mile or two up the trail I find a bag of Skittles in the front pocket of my running vest and choose an orange candy to pop into my mouth. Just in front of me, a dead branch falls from a tree. It clips a small rock, sending it hurtling down the hillside. After a moment I hear the rock plunk down in a tiny creek far below. I imagine it creating a small pattern of ripples, modifying creek's flow ever so slightly.

In the physical, butterfly-effect world, determinism is easy to see: branch hits rock, rock falls into stream, stream flows a little differently. Over thousands of years, after millions of tiny rocks plunk down in the stream and make their own tiny modifications, this stream might change its course altogether, exterminating whole ecosystems of life in the areas it abandoned, while giving life to new ones. Causes and effects seem to rule the natural world.

But what about my own actions? Are those determined too?

Intuitively, the answer seems to be *no*. Take my choice to eat an orange Skittle. I chose that Skittle in particular because I *wanted* to. Nothing external seems to have forced me to do it. My actions, it seems, are governed by me, not by some set of rules inscribed by the Big Bang 14 billion years ago. It's fair to say that I had *free will* to choose the orange Skittle.

But was that choice truly free?

I picked that Skittle for some set of reasons – because I was in the mood for a burst of artificially simulated orange-candy flavor, or because it stood out from the other, drabber Skittles, or maybe just because it was on top of the bag.

Now what if we follow one of those causal lines backwards. Why was I in the mood for the orange-flavored Skittle? Because I like the fruity flavor. Why do I like its fruity flavor? Because of the way my taste buds are arranged and connect to the sensory receptors in my brain, and because of my previous sensual experiences. All this history is encoded somehow in the atoms of my brain.

And why Skittles at all? Why didn't I choose a sip of water at that moment instead? I chose a Skittle due to the specific physiology of my body at that moment, which took into account my body's need for quick glucose, its evolutionarily encoded desire for sweetness, and perhaps the ploys of the clever marketing staff at Mars, Inc., who designed this product and its packaging to optimally play upon my psychological desires.

Each of these factors, in turn, has its own antecedent cause. Those marketing geniuses designed this Skittles package because of their own training and experience in marketing – which is itself encoded somewhere in the atoms of *their* brains.

I think about less trivial examples. Why am I here, on this path, thinking about this book? Why did I quit my job six years ago to start my own consulting practice? Why did I marry Lisa seven years before that? Why did I move to Boulder in the first place, allowing all these other events to happen? These are all things I wanted to do, but they were all determined because each has a chain of causation, drifting back into the far reaches of time. I married Lisa because I "knew" she was the love of my life. But why? Because of how we communicated. Because she liked to drink cocktails with me. Because I liked the way her legs looked in a short dress. And why did I like all those things? Ultimately, because of the status of all the atoms in the natural world, together with the laws of nature that govern how these elements interact.

Determinism, like its cousin, naturalism, sometimes looks cold and heartless – the polar opposite of spiritual. How can a pre-ordered world have room for life, for spirit, for freedom?

It turns out that determinism is actually a *prerequisite* for spirituality. This is because a deterministic world is simply a world in which things are ordered and predictable. A deterministic world is sensible. Things happen for a reason, not capriciously or arbitrarily. We decide to act not because of some random, uncontrollable impulse untethered to anything about *us* or what we want, but because of a desire that is "stored" in the physical world.

Determinism has deep roots in the annals of philosophy. While some ancient Greeks, like Epicurus, tried to make space for free will, the Stoics who followed held a more rigid view of causal determinism: every event occurs because of some prior event. A Stoic of virtue amends his will to suit the world and remain "sick and yet happy, in peril and yet happy, dying and yet happy, in exile and happy, in disgrace and happy."

The Roman statesman and philosopher Cicero summed up the Stoic sensibility:

> By "fate," I mean what the Greeks call *heimarmen* – an ordering and sequence of causes, since it is the connexion of cause to cause which out of itself produces anything. Consequently nothing has happened which was not going to be, and likewise nothing is going to be of which nature does not contain causes working to bring that very thing about. This makes it intelligible that fate should be, not the "fate" of superstition, but that of physics, an everlasting cause of things – why past things happened, why present things are now happening, and why future things will be.

The church, of course, put its own spin on the determinism/free will debate. Humans had to have free will, which was granted by God, in order to exercise moral judgement, even if we lived in a world constrained – determined, as it were – by God's ultimate authority.

The scientific revolution produced a more rigorously empirical worldview, closely tied to the physical world. Isaac Newton's mathematical theory of motion seemed to reveal a deterministic natural world. "Everything proceeds mathematically," he said. "If someone could have a sufficient insight into the inner parts of things, and in addition had remembrance and intelligence enough to consider all the circumstances and take them into account, he would be a prophet and see the future in the present as in a mirror."

The philosophers who witnessed the power of the new scientific revolution – such as the British empiricists David Hume, John Locke, and George Berkeley – were determinists. They believed all of our actions had causes, even if our own character was one of those causes. Hume said, "'tis impossible to admit of any medium betwixt chance and an absolute necessity."

Scientific determinism was articulated precisely for the first time by the eighteenth-century French physicist Pierre Laplace. "We may regard the present state of the Universe as the effect of its past and the cause of its future," he wrote. "An intellect which at a certain moment would know all forces that set nature in motion, and all positions of all items of which nature is composed, if this intellect were also vast enough to submit these data to analysis, it would embrace in a single formula the movements of the greatest bodies of the Universe and those of the tiniest atom; for such an intellect nothing would be uncertain and the future just like the past would be present before its eyes." This "intellect" became known as LaPlace's demon, a sort of secular, all-knowing superintelligence.

The twentieth century brought a revolutionary challenge to determinism, through the unexpected agent of science itself. This was the mind-boggling and thoroughly counterintuitive discovery of quantum mechanics, which introduced the possibility that at some level, nature might be truly random. But just as advanced physics was upending our confidence in determinism, the science of human behavior was reinforcing it. In the emerging field of psychology, scientists were adapting the mechanistic view of the Universe to human behavior. These behaviorists, including Ivan Pavlov and B.F. Skinner, believed that all human behaviors were consequences essentially out of our control, the result of environmental stimuli and our history of reinforcement and punishment.

In many ways, modern science bolsters the behaviorists' deterministic perspective. Some of the most compelling new evidence comes from the field of neuropsychology. Starting in the late 1970s, the pioneering consciousness scientist Benjamin Libet conducted a fascinating series of experiments that seemed to show that our actions are predetermined even when we think we've consciously chosen them. Libet had study participants sit down in front of a timer with a moving needle. He then affixed electrodes to the participant's scalp and asked them to do some small motor activity, like pressing a button or raising a finger. Subjects were asked to note the location of the timer's needle at the very instant they were first aware of the desire to act.

During the experiment, researchers recorded subjects' brain wave patterns to determine the moment at which activity in the brain's motor cortex began. It turned out that the brain started wheeling through its preparations to lift the finger three hundred milliseconds before subjects reported wanting to conduct the action. The brain seems to have "decided" to lift its owner's finger before the "owner" had the conscious experience of making the decision.

Subsequent experiments have been able to identify even longer durations between the brain's actions and participants' subjective decisions. In 2008, researchers at the Max Planck Institute used brain scanners to predict subjects' decisions a full seven seconds before test subjects were even aware of making them. In one version of the 2008 study, when subjects' left hemispheres were stimulated electrically, the likelihood of right-handed subjects moving their right hand increased from 60% to 80%. Yet another study measured cortex activity directly and found that the activity of 256 neurons was sufficient to predict with 80 percent accuracy a person's decision to move a full 700 milliseconds before she became aware of it. Even more creepily, neuroscientists working with mice have been able to remove memories by weakening specific synapses in their brains, and even artificially implant false memories by directly stimulating individual nerve cells with electrodes.

"Your decisions are strongly prepared by brain activity," explained Jon-Dylan Haynes, one of the Planck Institute study's co-authors. "By the time consciousness kicks in, most of the work has already been done." The Danish popular science writer Tor Norretreanders calls this the "user illusion," implying that consciousness itself is merely an illusion, but our actions are actually controlled by non-conscious parts of the brain and our conscious mind is merely a spectator.

The German philosopher Thomas Metzinger calls this the Myth of Cognitive Agency. We believe that we are the active thinker of our thoughts, and that we think rationally and in a goal-directed manner. In fact, we spend much of our lives zoning out, our mind travelling at the whim of those thoughts, encoded as physical patterns in our brains. We believe our thoughts follow us, but in reality, we follow our thoughts.

The enigma of free will

There's a problem with all of this, a problem so immediate and obvious it scarcely needs explaining: the simple fact of free will. I have the raw experience of making choices, a phenomenological fact that seems self-evident. Some ardent determinists, like Sam Harris, have denied the existence of free will altogether, calling it a mere illusion. But this doesn't seem to capture the "realness" of our ability to make a choice. Free will is a story that captures something essential about the experience of being human. To say that it's an illusion doesn't really add anything of value to this story. To put it another way: What would our experience be like if free will was *not* an illusion? Would it feel any different?

The world facilitates *both* determinism and free will. This is the very conundrum in which philosophers have become entangled for millennia, though some have offered solutions that attempt to accommodate both truths. One such solution is the "two-stage" theory of free will, originally proposed by the great philosopher William James in the late nineteenth century. James suggested that chance gives us a set of random alternatives, from which we make a single choice that turns the ambiguous future into a determined past.

The ability to make a conscious choice even when that choice is constrained by the natural world is sometimes called "free won't." The subconscious (deterministic) mind sends to consciousness a selection of choices, while the conscious mind selects from among them. In the example of the Libet experiments, the subconscious reports on the desire to raise a finger, but the participant consciously decides whether and when to honor that desire. We are free to do what we *want*, but not to do what we *want to want*.

Unfortunately, "free won't" doesn't really seem to solve the determinism/free will debate; it just defines the problem more narrowly. Even if we admit that subconscious processes are not governed through free will, we're still left wondering how the conscious mind selects freely among choices, without reference to the natural world. After all, this selection, and the subsequent real-world activity, are all governed by the deterministic laws of nature.

Compatibilism

We seem to have arrived at a paradox: free will exists, but determinism must exist too. Before we step forward, let's review how we got here:

1. Science and observation convince us that the natural world is the only world.
2. The natural world is ordered and governed by rules that allow us to explain and predict our observations.
3. Human behavior is rooted in the natural world, and is thus subject to its laws.
4. Our own choices, therefore, are determined by the state of the natural world.
5. However, we also have the direct experience of free will.
6. Somehow, determinism and free will must cohabit together. But how?

The main contemporary philosophical approach to unite determinism and free will is called *compatibilism*. Early compatibilists, such as David Hume and Thomas Hobbes, believed that we have freedom of action as long as we are not being coerced by some external agent. The contemporary philosopher Daniel Dennett believes that an action can be freely chosen, even if there are no alterative possibilities (i.e., our action is determined).

This gets at an important point: a "choice" is meaningless *unless* it is determined. Imagine, for example, that the random decay of atoms affected whether we chose, say, whether or not to eat vanilla or chocolate ice cream, or whether to marry our high school sweetheart. This would not be a choice in any meaningful way – it would be an arbitrary, externally-imposed condition of the physical world. To belabor this crucially important point, determinism is *required* for choice.

If I step up to the Ben and Jerry's counter and choose Cherry Garcia ice cream over Chunky Monkey, I am drawing on the stored information in my brain as well as the sensory input arriving in the moment. My choice can be seen as a logical pattern determined by physical information.

For example, imagine that I am perfectly on the fence about which ice cream flavor to choose, and just before I announce my order a random quantum process causes a single particle to materialize in my brain and somehow tip the balance toward Cherry Garcia. One may argue that me selecting Cherry Garcia is not "determined," in that it is not uniquely predictable from the state of the world before the particle materialized. But it is still determined by the external physical laws of the Universe. There is still no sense in which I have made a meaningful choice about what I "want." The choice was made *for* me, but by a random process rather than a chronologically deterministic process.

Physical events which are somehow caused by a random process such as quantum mechanics are in no way a matter of "our will." They are simply random. A choice must have a logical antecedent. Otherwise there is no basis for us to "want" anything.

Compatibilism has garnered its share of critics. William James called it "soft determinism" and thought compatibilists were creating a "quagmire of evasion" and were really just trying to mask their underling determinism. Immanuel Kant called it a "wretched subterfuge" and "word jugglery."

Some contemporary determinists have criticized compatibilism for exactly the opposite reason: that it is trying to sneak in free will. The evolutionary biologist Jerry Coyne, for example, thinks that scientific evidence does not show evidence of a mind separate from the physical brain, so choices are in principle predictable by the laws of nature. Compatibilism is a "semantic game," just philosophers playing around with words to obscure the fact that choice is simply an illusion.

Sam Harris, a determinist and opponent of compatibilism, believes that the popular conception of free will rests on two assumptions: (1) that each of us could have behaved differently in the past; and (2) that we are the conscious source of most of our thoughts and actions in the present. The first assumption is wrong, he says, because all our behaviors are influenced by the status of the world before us, and the second assumption is wrong because our thoughts and actions are created by the physical world.

So compatibilism, at least as it's been framed, doesn't seem to resolve that underlying tension between the fact of determinism and the visceral experience of free choice. Compatibilism's challengers, however, are defending one half of the equation or the other, but not really acknowledging that both determinism and free will *must* exist together.

Something, it seems, is blocking the way – something that has been wholly overlooked by the canon of Western philosophers. To really figure out how determinism and free will are connected – and to truly make progress on our scientific seeker's journey – we're going to need to find this hidden demon, and then we're going to extract it like a rotten tooth.

As it turns out, we've come across this demon before: it is called dualism.

I crest the ridge near Mt. Belford and a huge swath of the central Rockies comes into view. I'm at about 12,000 feet now, well above treeline, and the stubby aspens and pines have given away to prickly krummholz, and finally to a fragile carpet of alpine moss and wildflowers. I've just got a short ridge to ascend until I'm finally standing on the summit.

I take a quiet moment to take in the expanse, eating a crumbled piece of a granola bar– the Skittles, having contributed more than their nutritional and philosophical share, are long since gone. I do the standard circle check on my physical status: muscles a little tired but still feeling strong, lungs working well, hands still warm. A grayish bulk of clouds lingers on the horizon, but I ought to have enough time to summit and get back down to treeline before the afternoon turns threatening.

And so off I charge for the final climb. This part of the journey is steeper, and I'm no longer even pretending to run. I take short, determined strides, power hiking up the narrow trail. The air thins in my lungs and I narrow my focus, looking down, advancing one small step at a time. Eventually the summit will come.

As my calves hack away at the final ascent of the mountain, my mind hacks away at this strange philosophical conundrum. Compatibilism seems like the only way to resolve determinism and free will, but there is something *wrong* with it. I just can't quite put my finger on it. Something doesn't connect, doesn't quite feel right.

Part of my unease has to do with that semantic trickery James and Coyne were talking about. I'm all for spending lots of time in coffeeshops writing dissertations on academic arcana – I've done plenty of that myself – but maybe something's getting lost in the overly academic logic games. What might it mean to be free to do what we "want" but not what we "want to want," anyway? Is there really a meaningful difference between the various philosophical creeds and camps, such as soft determinists, event-causal indeterminists, Valerian determinists, impossibilists, illusionists, and daring soft libertarians?

The philosopher of science John Searle observed, "The persistence of the free will problem in philosophy seems to me something of a scandal. After all these centuries of writing about free will, it does not seem to me that we have made very much progress." So how do we make progress? What's missing?

Maybe something about these mountains will help. Maybe there is a way to get this philosophical debate off the page, so to speak, and into my heart, into the fire of existence.

I finally make it to the crest of the long ridge that connects Belford to its 14'er cousin, Mt. Oxford. Running at 14,000 feet leaves no room for mental meandering. It does for your mind and body what a lump of straight horseradish does for your sinuses at a Passover seder. It is an act of discipline that offers a clear mind and body as its reward. Whatever insights manage to filter through come in pure, uncomplicated breaths.

I double my breathing rate, one in and one out per stride. Determinism on the in-breath. Free will on the out-breath. How could we have both? A paradox.

One in, one out.

The spiritual seeker's version of a philosophical paradox is the *koan*. In the Zen tradition, a koan is a statement that appears contradictory on the surface, but illuminates something profound underneath. When we crack the *koan* by identifying an illusory assumption, we find the unified truth underneath.

The *koan* encourages us to hold two seemingly incompatible ideas in our head at the same time without forcing one of them to "win." The type of unity has always been a struggle for Western philosophers, who have tended to frame the key philosophical conundrums dualistically. We are either good or evil. We are physical or we are spiritual. There is a god or there isn't one. Either we have free will, or the world is deterministic.

It's dualism, I think. *That's the problem. That's the hidden demon.* I know right away I've hooked something, but it'll take some mental calisthenics to sort out quite what it is – sort of like jiggling a Boggle board until all the letter cubes settle in place.

Compatibilists think they've extracted dualism already, but they haven't gotten to the core of the matter. We can see this by stating the familiar positions of the main camps in a different way: Determinists believe that the world acts upon us. Libertarians believe that we act upon the world. Compatibilists believe both: that the world acts upon us and that we act upon the world.

All of these conceptions involve two separate things: *us* and *the world*. They have framed the problem as a relationship *between* two separate entities. Does the natural world control my thoughts and desires, or do my own desires have agency over the natural world? Or are both somehow true at the same time?

To take the next step, we need to define more closely what we mean by the "natural world." We really mean the ordered and predictable patterns that create a sensible external reality. The natural world is the laws of nature that generate the story we sometimes call the physical or material world.

But what if there is *no difference* between "me" and those laws of nature? What if they are literally the *same* thing, and it's all those gloomy European men in beards, unwittingly steeped in their own form of dualism, who invented the paradox in the first place? That would mean that human consciousness and the experience of agency – what we call free will – must be an *aspect* of the laws of nature. We are not *acted upon* by the laws of nature, or do we *act upon* them. We *are* them.

Think of it this way: If the laws of nature themselves – the very patterns and rules that govern the Universe – could experience awareness, what would it feel like? It would feel like being a human. Consciousness is what happens when the laws of nature start to be aware of themselves. This is the "mystical" part of mystical naturalism. Somehow the laws of nature are invested with self-awareness, and those laws can experience happiness and pain, meaning and desire. The laws themselves have become "alive."

I try sitting with this idea for a moment. We *are* the laws of nature, made conscious. It's meaningless to ask if the world acts upon us, or if we act upon the world. To put it in terms any child of the 1980's will understand – we *are* the world. When I choose an orange Skittle, it is identical to saying that the laws of nature choose an orange Skittle. True to determinism, the cogs of order and causality are what are choosing the Skittle – but true to free will, *I* am choosing the Skittle – as *I* am the cogs. Free will is merely a synonym for awareness itself.

As this idea settles in my mind, it comes to seem self-evident, as if in plain sight the whole time. *We are the laws of nature, made conscious.* Of course I make choices, just as a computer chooses how to calculate the answer to an arithmetic problem, just as the Earth chooses to rotate around the Sun, all according to the holy order of the cosmos we call the laws of nature. I am a self-aware node of the intricate cosmic web. And that web, of which I am a tiny part, created everything, from the first atom to this thunderous mountain to my own young daughters, two new nodes birthed from the matrix's womb.

CHAPTER 7

Why Does The World Exist?

Doubt can only exist where a question exists, a
question only where an answer exists, and an answer
only where something can be said.

- Ludwig Wittgenstein, Tractatus Logico-
Philosophicus

Mt. Belford's summit is broader than many others in the
Rockies, an accommodating swatch of flattish ground
accessed through a final climb over a cap of sharp
boulders. I finally make it to the top and take in my surroundings. I
haven't been on top of a 14'er since I met Lisa, fourteen years ago,
and while I'd intellectually stored the experience somewhere, I'd lost
its immediate emotional resonance. From the top of a 14,000-foot
peak, you are above everything. All the other mountain tops, all the
smaller hills and knobs, all the rivers and valleys and towns and glades:
they are all below you.

This little hook that seems to defuse the whole
determinism/free-will debate, this demon of duality, seems deeply,
profoundly true. But it's hard to really wrap my mind around it. My
brain, like everyone's, has evolved to perceive the world dualistically.
Our culture, steeped in Western philosophy and monotheistic
religious teachings, ceaselessly reinforces this illusion. Trying to
understand intellectually what this all means is like trying to picture a
four-dimensional object: my brain just doesn't want to go there.

I look out across range after range of mountains, letting all my highfalutin thoughts disperse into the thin, clean air in all directions. There's the Collegiate Range, a line of peaks stacked up towards the south, and there are the pointy Sawatches to the north. I can see the gray plume of a forest fire fifty miles away to the east, and the graying bulk of a thunderhead looming to the west.

Gradually, as my mind empties, the intellectual, philosophized idea – *I am the laws of nature, made conscious* – starts to translate into something more immediate, something I can *feel*. I imagine a three-dimensional grid spanning across the mountains, up into the blue sky above, down into the granite rocks below, curving and twisting and occasionally gathering together into dense, complex patterns – the scattered nodes of awareness we call humans. I imagine this web emanating outward from my fingers like Spiderman. I am part of its rhythm, part of the dance of life, of all existence.

Eventually, the realities of the prosaic physical world return, as those graphite stormclouds darken and start spattering dark circles of rain on the white rock. I've spent enough time in the mountains to have developed a sixth sense about weather, a formula that draws in the color of the clouds, the presence of lightning off to the horizon, and the distance back to treeline.

It's time to head down.

I keep a steady and careful rhythm back down the trail, fast enough to beat the weather, slow enough to avoid tripping over a root or misplaced stone and eating a piece of the muddy trail. Soon enough I'm safely below treeline and I settle into a more relaxed pace down through the forest.

I'm well underway on my scientific seeker's journey now. I've seen that the natural world is the *whole* world, and that it is distinguished by a profound order. It is this order that invests science and reason with their unique power. I've wrestled with the age-old philosophical debate of free will versus determinism, and seen that it's based on the Western illusion of dualism. The laws of nature do not act upon our consciousness, nor does our consciousness act upon the laws of nature. Rather, consciousness is a property of those very laws, much as the color red is a property of an apple. The universe itself is aware, and we are its eyes and ears.

What I still cannot fathom might be the most profound question of all: Why? Why does the world exist? Does it have some purpose? Must it exist, or is its very existence arbitrary? And can we ever really explain or comprehend the brute fact of existence itself? Does it have some understandable logic or purpose, or must it remain forever outside of our grasp?

One place to start exploring *why* the world exists is to ask *how* it came into existence. To answer *that* question, let's start with the origins of the known universe: the Big Bang.

The Big Bang

The term "Big Bang" was coined in 1949 by astronomer Fred Hoyle, who used the term derisively, preferring instead an eternal model in which the Universe simply exists, has existed, and will continue to exist, in more or less its present state. The Big Bang model remained merely speculation for another two decades, until a warm spring evening in 1964. That's when two astronomers, Robert Wilson and Arno Penzias, were testing a radio telescope on a New Jersey hilltop. Wherever they pointed their telescope, they found a constant level of excess noise. This didn't make sense, as they knew that space is not perfectly uniform – there are galaxies and star clusters here, gaps and darkness there. After ruling out some initial suspects, such as radio noise from New York City or after-effects of a nuclear bomb test several years earlier, they identified the likely culprit. On inspecting their antenna, they found that two pigeons had set up house inside it. Most likely it was pigeon poop that was mucking up their signal.

After they trapped the birds (the traps are now on display at the Smithsonian's Air and Space Museum) and cleaned the antenna, the strange pattern continued. Ultimately, Wilson and Penzias realized they were listening to a signal from the very beginning of the Universe, billions of years ago. The Big Bang theory had its first empirical confirmation.

In the years since, we've refined our understanding of the very early moments of the Universe. According to the latest and most comprehensive theories, the Universe came into being about 13.8 billion years ago. It emerged from a singularity, a mysterious point of infinite density and infinite heat. About 10^{-43} seconds later – an incomprehensibly small sliver of a second – the force of gravity separated from the other fundamental forces, and the elementary particles were created. The tiny new universe expanded at mind-boggling velocity until by 10^{-32} seconds, it was about the size of a grapefruit.

By the time the Universe was one second old, it was a sea of fundamental particles – neutrons, protons, electrons, positrons, photons, and neutrinos – with a temperature of about 10 billion degrees Fahrenheit. (The last of these, the tiny neutrino, is so small that any given one has likely never collided with another particle in its billions of years of life).

Over the next 380,000 years, electrons and nuclei combined to produce atoms, allowing light to radiate. This is the "afterglow" of the Big Bang that was detected on that spring evening in New Jersey. (Amazingly, you can detect this background radiation yourself: if you turn on your TV and see snow, about 10% of that signal is from the very birth of the Universe.)

Several hundred million years later, clouds of cosmic gas began to collapse under their own gravity, becoming hot enough to trigger nuclear fusion reactions between hydrogen atoms. The first stars were born, and these became crucibles that created the heavier elements. About nine billion years after the Big Bang, the solar system began to form, and with it the embryonic Earth.

The Big Bang is not controversial today among scientists, but when it was proposed many found it suspect for its supposed religious overtones. The prevailing "steady state" theory proposed an eternal universe, which could be interpreted as having no need for a creator.

The Big Bang theory seemed to support the Biblical concept of a finite universe created at a fixed point in the past; its originator, Georges Lemaitre, was a Roman Catholic priest. Stephen Hawking explains, "so long as the Universe had a beginning, we could suppose it had a creator. But if the Universe is completely self-contained, having no boundary or edge, it would have neither beginning nor end: it would simply be. What place, then, for a creator?"

The notion that the Universe began at a fixed point in the past seemed to challenge the non-religious. The physicist David Bohm called developers of the Big Bang theory "scientists who effectively turn traitor to science, and discard scientific facts to reach conclusions that are convenient to the Catholic Church." Otto Heckman, a physicist studying cosmic expansion, observed that "some younger scientists were so upset by these theological trends that they resolved simply to block their cosmological source." The English astronomer Sir Arthur Eddington wrote that "the notion of a beginning is repugnant to me ... I simply do not believe that the present order of things started off with a bang ... the expanding universe is preposterous ... incredible ... it leaves me cold." Einstein, too, was troubled by the conclusion that the Universe began at some point in time, though ultimately he came to embrace the theory; not long before his death in 1955, he referred to his initial doubts about the Big Bang as "the greatest blunder of my career."

The religious, on the other hand, embraced the theory. Pope Pius XII, at a conference in the Vatican in 1951, declared that this new theory of cosmic origins bore witness "to that primordial *Fiat lux* uttered at the moment when, along with matter, there burst forth from nothing a sea of light and radiation ... Hence, creation took place in time, therefore there is a creator, therefore God exists!"

As it turns out, the Big Bang theory has little bearing on questions about whether God created the Universe. It tells with impressive detail the origins of our universe, but it doesn't answer the deeper question: why anything exists in the first place. In fact, it raises more questions than it answers. Where did the Big Bang itself come from? What forces brought it into being? Is our universe part of a greater multiverse? If so, where did that multiverse come from?

To complicate things further, the Big Bang did not occur at a given point in time and space – it *created* time and space. According to the prevailing Hartle-Hawking model, the beginning of the Universe was not a clear "bang," but rather can be imagined as a cone from which time emerged.

Where did this cone itself come from? As best we can tell, it appears to have emerged from nothing at all. It's difficult to conceive of this nothingness. It's not the nothingness of an empty space, or of a vacuum; it's the absence of space itself.

The physicist Alex Vilenkin offers a physical description of the state of nothingness that could have birthed the Big Bang. Imagine spacetime as the surface of the sphere. Now imagine the radius of the sphere shrinking to nothing. The surface of the sphere disappears completely, along with spacetime itself. We're left with a "closed spacetime of zero radius." Vilenkin thinks that this state of nothingness — a bit of energy-filled vacuum one hundred-trillionth of a centimeter – could birth itself into the somethingness of the Big Bang, through the laws of math and physics.

This might appear to show that the world needs no creator; math can do all the work of creation by itself. But it still does not tell us where the potential for creation came from. From where did the mathematical laws that birthed the Big Bang arise? What is this "closed spacetime"? It is not truly nothing. It is *something*.

The distinction between the physical concept of "nothing" and the philosophical one is subtle but important. It is impossible to talk about *nothing* without it being a subject or object in our sentence, thus giving it form and essence. A joke illustrates this point:

- Premise 1: Nothing is greater than God.
- Premise 2: A ham sandwich is better than nothing.
- Conclusion: A ham sandwich is greater than God.

The problem is that the term "nothing" means something different in Premise 1 and Premise 2. In Premise 1, it refers to an imagined entity; while in Premise 2, it refers to the absence of something. Ultimate nothingness, the kind hinted at in Premise 2, would mean nothing has or will ever exist – no objects, no thoughts, no ideas, no experiences, not even an empty void, for a void has some essence. Real nothingness has no definition, no existence, no meaning.

The fact is, even before the Big Bang, there was *something*: a possibility, a pregnant cosmic womb. Where did that possibility come from, and who invested it with the ability to birth a universe?

I'd like to explore two main types of responses to this question. The *teleological* approach posits that the world has an ultimate purpose; it is oriented to some goal. The second approach, which we might call the *logical* approach, is that the world has no ultimate purpose and requires no intelligent creator; the rules of logic make the existence of the Universe necessary in a way that requires no further explanation.

The teleological explanation

For the religious, the teleological approach needs little explanation: God (or a combination of deities) created the Universe, and that's that. This is St. Thomas Aquinas's idea of God as "ex nihilo": a timeless and necessary being who creates the Universe as an act of free choice. As scientific seekers, though, we need to dig a bit deeper.

From the standpoint of science, one of the most compelling teleological arguments hinges on what's known as the "Goldilocks Enigma," in the coining of physicist Paul Davies. Cosmologists have demonstrated that the settings of the Universe are "just so" to allow for the creation of life. This "Goldilocks" universe is unreasonably suited for life, far beyond the bounds of coincidence. Davies offers this example: a neutron weighs exactly 1.00138 times as much as a proton. If this seemingly arbitrary figure was even a tiny bit smaller, atoms would collapse upon themselves, molecules would not have been formed, and chemistry itself would not exist. There are dozens of other examples, all of which seem to have conspired to create a universe that sustains life.

Our existence here on Earth is equally amazing. Our planet had to be just the right distance away from our sun to generate life, and it had to have the right sort of moon to draw away asteroids and protect its growing biosphere. The climate had to be stable enough for years to enable life to become more complex. A comet had to hit at just the right time to wipe out the dinosaurs and allow our ancestors to thrive. Environmental pressures and competition had to be just so in order to allow our brains to evolve self-awareness.

The Goldilocks Enigma seems to show the unmistakable hand of a Creator. How else could our universe be so perfectly tuned to develop life? It turns out, however, that the apparent fine-tuning of our observable universe is far from conclusive proof of a cosmic design. The most compelling reason for this is that ours may not be the only universe. There may be many universes out there, perhaps an infinite number: a multiverse. One particularly exotic description of the Multiverse is that our entire universe exists on a brane that floats in a higher dimension called the "bulk." Every few trillion years, the branes collide and set of Big Bangs, birthing new universes each time.

Each universe within the Multiverse may contain a random assortment of initial settings. Together, the Multiverse would contain every possible, logically coherent setting. In this reality, the only creatures who can observe a universe are those that live in a universe that gives rise to conscious observers. It's no coincidence, therefore, that the universe we perceive is one that seems fine-tuned to create us. This logical truism is known as the anthropic principle.

There is nothing in science that tells us the multiverse theory is wrong. However, it may be equally impossible to confirm it, even in principle. If other universes are disconnected from ours and we cannot communicate with them, it would be impossible to ever know if they exist.

A potential problem with the Multiverse is that it violates the principle of Occam's Razor, the idea that the simplest explanation is usually the best. On its face, the Multiverse seems needlessly complex, adding uncountable universes to an already unfathomably complex reality. On the other hand, a multiverse may best satisfy Occam's Razor, because infinite possibilities are actually simpler than a single specific and seemingly arbitrary one.

But all of this is a bit beside the point. Let's accept, for argument's sake, that the Multiverse is real: There are an infinite number of universes, and naturally we live in one that supports self-aware life. That same nagging question remains: where did the Multiverse come from? Who, or what, endowed Creation with the power to make a multiverse capable of making life-spawning universes?

This is the famous *turtles* problem. As the story goes, the philosopher and scientist William James, after presenting a lecture on cosmology, is approached by a little old lady. "This is all very interesting," she says, "but the Earth is actually supported by a giant turtle."

"But on what does that turtle stand," James asks patiently.

"Why, on a *second* turtle," the little old lady answers.

"And what in the world does the second turtle stand on," James asks, hiding his indignance.

The old lady smiles victoriously. "It's turtles all the way down!"

To some philosophers, the so-called turtles problem is enough to assert that God exists. The German polymath Gottfried Wilhelm Leibniz called this the "Principle of Sufficient Reason." Every question has an answer, he believed, and our first question will always be: Why is there something rather than nothing? Implicit in this idea is the premise that everything has a cause. If we trace the cause of everything in the Universe back through time, we will eventually get to a first cause. This is what we call God. God created the world through free choice, motivated by his infinite Goodness.

Many contemporary scientists also see the fingerprints of an underlying intelligence in the world. The scientist Michio Kaku says, "I have concluded that we are in a world made by rules created by an intelligence. To me it is clear that we exist in a plan which is governed by rules that were created, shaped by a universal intelligence and not by chance." Robert Lanza, the doctor and philosopher of science, calls the seeming tilt of the Universe towards generating life *biocentrism*. The idea is that consciousness and life are not just incidental to the Universe, but that the Universe was specifically designed for them.

The logical explanation

The teleological argument has become the third rail among modern rationalists. Even to suggest that the world may have a purpose sounds too much like the debunked pseudoscience of Intelligent Design.

Evolution is a particular source of teleological queasiness among scientists. Evolution itself does not have desire or intentions – it is simply a mathematical process that through its own logic happens to generate organisms that are better adapted to their environments. Stephen Jay Gould compared evolution to a drunk guy leaning against a wall, who inadvertently slips and finds himself in the gutter. He didn't aim to be in the gutter; he's just there because the wall was a boundary that changed his direction. Evolution does not proceed in a particular direction; it just happens to head towards complexity because of environmental pressure.

When Darwin travelled to the Galapagos, he saw that finches on different islands had different types of beaks, each adapted to the trees on its own island. There was no master plan commanding beaks to develop in a particular way. Rather, the finches with less effective beaks were less likely to reproduce, and those with random mutations that enhanced their ability to find food were more likely to reproduce and convey their gene pool to the next generation.

The process of evolution, naturalists say, does just fine on its own without the need for an invisible hand to guide it. The fact that evolution seemed "designed" to create us advanced thinkers is another case of the anthropomorphic principle: we are wowed by the miracle of existence because we happen to be a species that evolved that capacity. On this reading, consciousness and higher intelligence are not things that evolution is driving *towards*, but accidental features that just happened to crop up in one of the Earth's millions of species. And in fact, evolution does not lead inexorably to greater complexity and self-awareness; the dinosaurs flourished for several hundred million years without becoming any smarter. (As proof, see the famous *Far Side* cartoon depicting several ne'er-do-well dinosaurs smoking cigarettes, captioned: "The real reason dinosaurs became extinct.")

Some scientists propose that evolution may follow some underlying path called an "attractor." Repeated patterns in evolution, such as the eye, which has developed independently dozens of times, could be reflecting some such hidden attractor. This is a purely mathematical concept, requiring no intelligent designer *per se* – unless, of course, mathematics *itself* is the intelligent designer.

Another logic-based explanation for the existence of the world uses Occam's Razor. If *something* is simpler than *nothing*, then perhaps no further explanation is needed for the world's existence. This is along the lines taken by the philosopher Robert Nozick. In his libertarian classic *Anarchy, State,* and *Utopia,* he posited a "principle of fecundity" in which all possible worlds exist simultaneously. When people complained that they found this view strange, he responded "Someone who proposes a non-strange answer shows he didn't understand the question."

Finally, we have Derek Parfit's cosmic selector, some special feature that chooses what a universe will look like. Maybe the Cosmos favors the simplest reality, à la Occam's Razor, or maybe it wants the reality with the most good, or the reality that is the most beautiful. Perhaps it's one of these selectors that logically compels the Cosmos to exist. The universe might have to exist because it is good to do so, or that it is simpler to exist than not to exist.

Uniting the teleological and logical arguments

Although the teleological argument and the logical argument seem to be diametrically opposed, this is a semantic illusion. We may define the Creator as God or as logic, but the conclusion is the same: the world exists because it must.

We are at the doorstep of a deeply profound realization: *The theological and the naturalistic explanations for the existence of the world both lead to the same place.*

We can explore this a little further by examining two logical terms often used to argue for or against a Creator: *necessary* and *contingent*. Things that are necessary simply must exist, with no other antecedent cause or explanation. Things that are contingent owe their existence to some other causal factor. A contingent thing *could have been otherwise*, had the world been just a bit different. Me winning Powerball is contingent upon my buying a ticket, and on a set of six Ping Pong balls with those exact numbers being selected in the drawing. My eating an orange Skittle is contingent upon the arrangement of the Skittles in the bag, my hunger status as relayed to my brain by my biochemical systems, and so on.

If there is a logical explanation for the existence of the world— for example, that something is simpler than nothing and so Occam's Razor compels the world to exist — then we can say the world is *necessary*. If the existence of the world is necessary, then that logical force that brought it into existence is synonymous with God, the necessary ultimate Creator.

If, on the other hand, there is no logical explanation for the existence of the world – if it just *is* – then the existence of the world fulfills all the criteria of a miracle. It cannot be explained through the workings of the natural world and it has no cause. Because it exists, it *must* exist, and so it, too, is necessary. We are well within our bounds to label the miraculous force that brought this world into existence God.

The distinction between contingent and necessary events is just a convenience we have invented to distinguish where things appear in the chain of causation. The event that appears the furthest in the past we label "necessary," while those at other points on the time dimension we call "contingent."

In actuality, *everything* is necessary.

Contingent events are often defined as events that *could have been otherwise*. But for any event to have been otherwise would require a wholly different world. Even quantum events, which do not appear to be caused, are still *as they are*. We've got one world, and it must be just as it is.

The limits of science and reason

The world must exist because it *does* exist. But we seem no closer to truly understanding the ultimate question we started with: *Why?*

We've come to a boundary, a cliff's edge beyond which science and reason cannot venture. Science is immensely helpful in understanding the very early moments of the Universe, and everything since then. Before science, every culture had its own origin story. A Greek myth tells the story of Nyx, the Goddess of the Night, born of Chaos itself. The Arapahoe Indians, who lived in these very mountains through which I run, believed the world was created from mud retrieved by a turtle in order to appease a man looking for a place to set his pipe. And the Biblical creation myth tells the story of a deity creating heaven and earth, and then inventing a flawed man and an apple tree to live in it.

Science has allowed us to replace these myths with a clearer picture of how our universe developed, and our picture of earliest moments of the Universe continues to get clearer and clearer. But something curious happens as science progresses. The more we know, the more we *don't* know. Our knowledge of the world is like a flashlight illuminating a cave; as we continue to explore deeper into the cave, the tunnel continues out of sight. We uncover one turtle only to find another turtle underneath.

We can't possibly know why we're here and why the Universe exists (if there even is a "why"), instead of nothing at all. But while we may not know *why* we're here, we do have the certain knowledge of existence itself. We can still marvel at this deepest existential riddle, and inhabit the miracle that I am here, and you are here, and we are able to experience our own creation.

CHAPTER 8

What is God?

I believe in God, but not as one thing, not as an old man in the sky. I believe that what people call God is something in all of us. I believe that what Jesus and Mohammed and Buddha and all the rest said was right. It's just that the translations have gone wrong.

— John Lennon

There's nothing quite like seeing the trailhead come into sight at the end of a long run. Today's route was especially grueling and that wooden sign marking the end of the trail is more welcome than ever. I trot past it and out into the parking lot and stop my watch, and my mind turns from the intricacies of abstract analytic philosophy to the more corporeal pleasures ahead: an air-conditioned car, a hot shower, a Himalayan buffet, and a frosty microbrew.

I'm staying tonight in the town of Leadville. The bed and breakfast I've found is billed as a hippie hotel, with a 70's style tie-died banner hanging out front. The place is a paean to pot, the newly legalized substance that is the source of an infinite number of Mile High jokes. But my sinful substance of choice is not weed, it's beer, so I indulge in a hot shower, scrub the trail dirt off my ankles, and head into town for a pint or two.

The first brewery I stumble upon is Periodic Brewing Company, recognizable by its logo, the black boldface Helvetica letters *Pb* on a square white background. Get it? Pb ... lead ... Leadville! I order a Sugarloaf Amber which, according to its label, "tastes like it smells when climbing Sugarloaf Mountain in the rain." A frosty mug is soon emptied, and I order another as I watch the mountain crowds file in and out. The combination of a hard day's running, a cold beer, and a cozy middle-of-nowhere bar give me a feeling of being *here*, being lost but at home as I watch the skyline of the Collegiate Peaks flatten and fade in the thin light of dusk.

Ever since college, a bit too much beer and a bit too little food has put me in a philosophical mood, and I pay my bill and wander out into the cool Leadville evening. A strip of hipster bistros and quirky coffeeshops populate the newly gentrified main drag, just as it does in many formerly lost Colorado mountain towns. Leadville's been slow to join the party, and even a block from the strip the town is silent. I wander a few blocks this way and that, strolling through the grid of simple houses, and soon I reach a bluff that marks the edge of town.

As the moonlight glints off the mountains in the distance, I feel that uncanniness you can only get outside, in the mountains, at night: a sense of being simultaneously tiny and part of something huge, witnessing the mystery and being part of it too. Standing out here, with a few pints of Sugarloaf Amber in my system, watching a silvery wisp of cloud pass in front of the moon, I feel a connection. But with *what?*

God, I guess. But what is God, if it even exists? Can science help us answer that question, or is it another of those ineffable, eternal mysteries? Can we define God? Or is God, by definition, *beyond* definition?

In much Western philosophy, it's taken as given that God refers to the all-powerful, all-knowing deity of monotheism. This is the God, almost invariably male and white, who strokes his long beard and points his fingers down from the clouds. We don't need to spend too much time evaluating that specifically Biblical conception of God here; we'll come back to Him in the next chapter on religion. Instead, let's start by examining the evidence for a more abstract, less concrete conception of God: an all-powerful, all-good, supreme being.

One way to come to a belief that God exists is to declare it as an article of faith, a raw, brute fact we choose to simply accept on its own grounds. Alvin Plantinga, a Notre Dame philosopher who is a central figure in theistic philosophy, argues that evidence should not be a necessary requirement for belief. We believe in other people's minds, after all, without direct evidence of them. We believe in the past. We are just as warranted in believing in God.

As scientific seekers, we know that this won't fly. For one, it violates that sacred principle of the Universe: that it is ordered and rational. If we want to understand the ultimate truth of the Universe – and therefore God – we must speak its language. That is to say, a conception of God must at least be consonant with reason, even if a true proof is beyond the reach of rationality.

At first blush, this may seem to tilt the playing field against God before the game has even started. But while the very idea of God may seem beyond the reach of logical arguments, throughout history many incisive thinkers have offered thoughtfully reasoned arguments for its existence, going back at least as far as the Bible itself. I've organized these into five categories: the *ontological* argument; the *cosmological* argument; the *design* argument; the *idealism* argument; and the *statistical* argument.

The ontological argument

The *ontological* argument follows a somewhat tortured logical path to make the case that God must exist. One of the earliest such approaches is the medieval Christian philosopher St. Anselm's "Perfect-Being" theology. God is the most perfect conceivable being, and it is more perfect to exist than to not exist. Thus, God *must* exist. St. Anselm's somewhat notorious proof has been criticized by many, among them Thomas Aquinas, who thought that it was impossible to even conceive of a transcendent God, and by Kant, who thought it confused the *idea* of God and God himself. The real dealbreaker is that the definition of "perfect" is in the eye of the beholder. Does a perfect being have emotions, or is it dispassionate? Does it create a world in which no people feel pain, or does the experience of pain somehow make the world more perfect? The devil, so to speak, is in these details.

The cosmological argument

The *cosmological* argument goes back to Aristotle and follows the logic of causation to its natural boundary. Everything must have a cause. But if we move backwards along the causal chain, back and back further, we eventually must get to an uncaused "first cause." This cause must be something outside of the line of causation altogether – something God-like.

The most sophisticated version of the cosmological argument was articulated by Leibniz in the seventeenth century. The universe, he thought, is contingent – that is, it might not have existed, and therefore, there must be a reason for its existence. For example, it could have been created by some other being.

But that being would require an explanation as well. Either this chain of causation – turtles on the backs of turtles – comes to an end, or it goes on *ad infinitum*. If it does come to an end, that end must be self-explanatory – that is, it must be necessary. If the chain of causation goes on ad infinitum, the entire chain must have been caused by something outside of it, and the existence of *that* being must be self-explanatory, and therefore necessary. Either way, says Leibniz, there must be a necessary cause to the Universe.

Leibniz goes a step further, ascribing some distinctly God-like attributes to this self-explanatory first cause. It must be eternal, and it must be infinitely powerful, because it caused the world to come into existence. It must also be infinitely intelligent, because it created a world in which intelligence exists. And it must be morally perfect, because in its infinite wisdom it can understand what is good.

Interestingly, under the lens of mystical naturalism, the cosmological argument generally stands up. Our typical evaluation of the God question from a scientific standpoint is dualistic; we imagine a God "out there," acting upon the Universe. If, instead, we remove the illusion of duality, we can imagine God as simply *everything*. God is the order and existence of the Universe itself, and all that comprises it. In this conception, God indeed is infinitely intelligent, because it contains within it all possible information. This God is also all powerful, comprising the rules and substance needed to make the Universe go.

Where the cosmological argument falters is when it becomes too concrete about ascribing morality to this God. Morality, as we'll see in the next chapter, is a human-defined concept, and what exactly it means is subject to human debate.

The design argument

The *design* argument points to certain features of the Universe, such as the fact that it's uniquely suited for the emergence of intelligent life, to argue that it must have had a cosmic designer. This argument is sometimes conflated with intelligent design, the scientifically discredited claim that the evolutionary process alone is not sufficient to create intelligent life, so evolution must not be an accurate account of how humans came to be. In fact, science is resoundingly supportive of evolution as the mechanism by which life as we know it emerged.

The association between the design argument in general, and the concept of intelligent design specifically, is unfortunate. While science can definitively confirm that the process of evolution gave rise to complex life, it simply can't weigh in either way on whether the world itself had a designer. The best we can do is wonder that it exists at all.

The idealism argument

A different tack altogether is *idealism*, the idea that the world is essentially comprised of spirit. It is typically seen in contrast to naturalism, which is (imprecisely, I think) associated with the idea that all is physical. Keith Ward, the Christian philosopher, puts forth this argument in its most detailed form. According to the idealists, the fact that there is consciousness is evidence itself of some sort of spiritual entity. Furthermore, this entity must have a moral charge, because all of us inherently have a shared moral sense. That sense must come from *somewhere*.

Unfortunately, there are a few logical flaws embedded within the idealism argument. First, it is based on a dualistic view of the physical and spiritual worlds. Why does a world made wholly of spirit preclude a world made of physical stuff? What if the physical stuff itself is the stuff of spirit?

The premise that consciousness is inherently moral is also problematic. Stating that we have a shared intuitive sense of morality is circular. For example, Ward says we would all intuitively understand that it is moral to save an infant from certain death from falling out a window. But the only way we can prove that this is a moral act is through logic and empirical observation, not through God-given intuition. In fact, it's exactly when we depart from rationality that morality becomes problematic. The list of religious laws that conflict with our rational moral intuitions is long.

Morality is a human construction, not a cosmic one. We don't, in fact, all have the same intuitive understanding of morality. While identifying the moral course of action in strawman cases like a falling infant seems obvious, it's not so straightforward in more controversial situations. Is abortion moral? Prostitution? What about contributing $1000 to your local art museum, instead of using that same money to buy malaria nets in Africa, where it can do far more good? There are no absolute answers to these questions; the best we can do is apply reason and science to them and make the best choices we can.

The statistical argument

A final argument for God leverages statistics to determine the likelihood a God exists, given our observations of the world. Oxford philosopher Richard Swinburne is one of the most well-known proponents of this type of proof. He argues that the existence of God is more probable than the non-existence of God, and therefore is a rational idea to hold. The total balance of evidence for God includes the existence of the Universe, its lawfulness, the history, and even the presence of evil. It is more likely than not, therefore, that there is a God.

Swinburne's argument is a cousin to the design argument, with some statistics thrown in give it a scientific veneer. The problem is that ascribing the characteristics of the Universe to God doesn't tell us much besides simply assigning a label to a bunch of observations. We're still left with the task of defining this God, so it doesn't seem we've ended up much further than where we started.

God and dualism

There's a bit in each of the arguments for God that seems compelling, but none seem to fully acknowledge their own logical limitations. Most critically, they fall prey to that trope of Western philosophy that continues to crop up on our journey: the demon of duality.

The dualistic conception of God is a relatively recent invention. The classical Greeks had, in many ways, a much more unified theistic understanding. Stoicism, a school of Hellenistic philosophy founded by Zeno of Citium in Athens in the third century BCE, held that there is only one substance, identified as God. In his own poetically stirring prose, Chrysippus, a prominent Stoic philosopher, wrote, "The Universe itself is God and the universal outpouring of its soul; it is this same world's guiding principle, operating in mind and reason, together with the common nature of things and the totality that embraces all existence; then the foreordained might and necessity of the future; then fire and the principle of aether; then those elements whose natural state is one of flux and transition, such as water, earth, and air; then the sun, the moon, the stars; and the universal existence in which all things are contained."

Aristotle, Plato, and Aristarchus had a more dualistic view: they believed in a dichotomy of mortals existing below with a parallel heavenly realm above inhabited by gods. In the centuries after the birth of Christ, the Gnostics and Manichaeans believed in the sharp distinction between body and soul, matter and spirit. Their deity had a split personality too: the spirit is created by the good god, but matter and bodies are the creation of the evil god.

As Christian monotheism spread, the Church became terrified of any threat to the idea of a maximally powerful, maximally good God that was quite separate from man. Anyone attempting to unite Classical thought with monotheistic dogma was venturing onto precarious ground, at risk of excommunication, or worse, death. This union was attempted most notably by St. Thomas Aquinas, who was inspired by Aristotle. He was condemned for his heresy, but ultimately hewed to traditional Christian thought when the two belief systems differed too greatly.

Spinoza's Deus sive Natura

No one, perhaps, suffered more in his attempt to extricate spirituality from the bounds of dualism than the seventeenth-century Jewish philosopher Baruch Spinoza. Born in 1632, Spinoza was a Sephardic Jew whose family immigrated from Portugal to Amsterdam. He was a lens grinder by trade – an occupation which may have cost him his life through glass inhalation – and lived a simple life. But his ideas were hugely controversial within his orthodox Jewish community. He was a pantheist, and saw objects in the physical world as attributes of God. To Spinoza, there was a deep order or rightness with the world. Human suffering was but a minor discord in a larger cosmic harmony.

Spinoza offered the idea that the world is *causa sui:* a cause in itself. He thought that all reality – physical and mental – consists of a single substance, *Deus sive Natura:* "God or Nature." God couldn't possibly stand apart from nature, because that would limit either God or nature. The world itself is divine and simply must be as it is. He believed that rationalism – an intellectual love of reality – is the highest possible calling for humans, and that when he thought, it was the same as God or nature thinking. To Spinoza, God, synonymous with nature, unfolds in an intelligent way, and all beings are manifestations of God.

In exchange for his bold insights, Spinoza was issued a extraordinary excommunication, or writ of cherem, by the Talmud Torah congregation of Amsterdam. It read, in part:

> The Lords of the ma'amad, having long known of the evil opinions and acts of Baruch de Espinoza, have endeavored by various means and promises, to turn him from his evil ways. But having failed to make him mend his wicked ways, and, on the contrary, daily receiving more and more serious information about the abominable heresies which he practiced and taught and about his monstrous deeds ... they have decided, with their consent, that the said Espinoza should be excommunicated and expelled from the people of Israel. By the decree of the angels, and by the command of the holy men, we excommunicate, expel, curse and damn Baruch de Espinoza ... Cursed be he by day and cursed be he by night; cursed be he when he lies down, and cursed be he when he rises up; cursed be he when he goes out, and cursed be he when he comes in. The Lord will not spare him; the anger and wrath of the Lord will rage against this man, and bring upon him all the

curses which are written in this book, and the Lord
will blot out his name from under heaven.

That's pretty tough language for someone expressing the
essential unity of the world, and perhaps an indication of how
dangerous the very idea of oneness is to religious fundamentalists.
The idea that God was everything – in contrast to the Jewish God of
the Torah – was a mortal threat to his synagogue's brittle religious
foundations.

Spinoza took his books and lenses and lived the life of a pure
philosopher until his death in 1677. His legacy is of a philosopher's
philosopher, someone who not only spoke of the life of the mind, but
lived it. "As a teacher of reality," wrote Howard Bloom, "he practiced
his own wisdom, and was surely one of the most exemplary human
beings ever to have lived." Einstein, when asked whether he believed
in God, said "I believe in Spinoza's God, who reveals himself in the
orderly harmony of what exists, not in a God who concerns himself
with the fates and actions of human beings."

It's in Spinoza that I find the closest parallels to the concept of
God I'm developing on my own journey. Spinoza's God is not only
understandable through rational means, it *is* rationality. It is not
separate from us, but a part of us, and we a part of it. Spinoza even
explicitly connects spirituality and nature: *Deus sive Natura*. He's just
about as close to a mystical naturalist as you can get, this Jewish
glassmaker who died three centuries before I was born.

Why believe in God?

So we're compelled to ask: if the traditional, dualistic conception
of God has so little evidentiary support, then why do people continue
to insist on believing in Him?

Non-theistic conceptions of God have always been met with
skepticism, and often outright hostility, from those in power. For
millennia religion has instilled within its subjects fear to keep them in
the fold – fear of people with different beliefs, fear of a nasty afterlife,

fear of falling out of grace with God. Fear, it seems, can be an even more powerful motivator than reason.

Beliefs in specific religious images of God don't always have to do with attempts to secure power and define in-groups and out-groups. A rather cynical, though plausible, explanation of flawed theistic logic is that humans are simply not good at critical thinking. Our brains, as we've seen, are not well suited to objective logical argument unless we train them accordingly. Without exposure to scientific and rationalist ways of thinking, we lack the tools to really evaluate truth claims.

Combine the coercive power of religion with the imperfect human brain and you're likely to stray far from the truth. Religions have an expert way of "back calculating" from their inscribed beliefs to build a logical proof for God. Coincidentally enough, each of these logical paths happen to lead directly to their own religion's doctrines.

As an analogy, consider the metaphor of epicycles, suggested by Pirsig in *Zen and the Art*. In ancient Rome, astronomical models assumed that Earth was the center of the solar system, and indeed the entire universe. The astronomer Ptolemy observed that these models failed to account for the motion of the planets. But rather than abandon the faulty Earth-centric model, he assumed that each planet moved on a small sphere, called an epicycle, which in turn moved on a larger sphere. Ptolemy could continue adding spheres to reflect the measurements of the motion of the planets and retain the Earth-centric model. Eventually, of course, the geocentric model could no longer be sustained, and the heliocentric (sun-centered) universe was born.

Statisticians have expressed the underlying problem in mathematical terms. A statistical model shows signs of "overfitting" when it contains too many parameters to support the data – when it is, in effect, "overthinking" the data. Theologians often do the same to prove God: they paste on more and more epicycles, more parameters, to protect their own specific notions of God. The eleventh-century Muslim theologian Al-Ghazali said that following the same formulaic path to religious truth will inevitably lead to the same conclusions, which happen to be the laws of Islam, such as fasting, charity, and praying towards God. Modern theologians like Alvin Plantinga and Keith Ward argue not only for God, but specifically for the monotheistic Christian God of the Bible. Ward believes that the existence of Christ as having been risen from the grave has a reasonable logical path: an overall creator is likely to exist, and it makes sense that this creator would have communicated to humans.

It's perilously easy to start with seemingly reasonable premises, and end up somewhere unreasonable, if not altogether absurd. One reason is that the veracity of a logical argument is very sensitive to small errors in reasoning along any part of the logical chain. These errors build upon each other in a subtle way, resulting in an irrational conclusion based on a series of apparently rational deductions. (To greatly oversimplify, imagine a logical chain with seven steps. Each step has only an 10% chance of being false – barely higher than the likelihood of drawing an ace out of a deck of cards at random. Multiply these "almost-perfect" steps together, and you're left with a conclusion that is more likely to be false than true.) Even a small amount of logical wobbliness anywhere in the system can land us somewhere wholly illogical.

Permit me to take a brief side journey to illustrate this point. Let's call now upon the Flying Spaghetti Monster, the tongue-in-cheek religious mascot cheekily proposed to oppose the teaching of intelligent design in schools. Our goal, for this exercise, is to prove through a series of intuitively obvious logical steps that the Flying Spaghetti Monster indeed is the deity that rules the Universe.

Consider this airtight argument:

1. Humans possess something spiritual that is distinct from the physical world, and that thing is awareness.
2. Our awareness is comprised of deeply felt passions. These passions must come from some sort of non-physical realm.
3. The universal experience of being human shows that we all have three spiritual experiences: 1) we are physically

incorporated and require sustenance from the material world; 2) we crave connection with the spirit realm; and 3) we fear that which is unknown and beyond our understanding. These three attributes of being human are so essential that they must reflect the three essential attributes of the deity.

4.   On Earth, which is the deity's creation, wheat is the most essential symbol of physical sustenance. We know this because it appears repeatedly in spiritual texts, from the Quran to the Bible. One attribute of the deity, therefore, must be a plentiful and universal form of wheat. The most iconic form of edible wheat is spaghetti.

5.   The craving for connection with the spirit realm has been associated throughout history with the desire to fly. We dream of flying. Our myths, from the Hindu Vimana to the Greek's Icarus, involve flying. The deity must therefore be able to fly.

6.   The deity is mysterious and fearsome. The concept we use to capture this attribute is that of a monster, from the Jewish dybbuk to the Islamic djinni.

7.   ***Conclusion:*** The deity therefore must be able to fly, must be made of spaghetti, and must be a monster. Therefore, it must be a flying spaghetti monster.

The point is that logical arguments, unless perfectly executed, are likely to yield a false sense of certainty about our theistic beliefs.

Atheism

If logic can't lead us to our traditional monotheistic conception of God, then perhaps there is no God at all. This brings us to the arguments for atheism.

One of the most obvious arguments for atheism uses Occam's razor. What is simpler: an imaginary, all-powerful, all-knowing deity that has never been observed and has no cause or explanation, or the naturalistic world we experience and understand? If science and logic explain things, why do we need to add a God? As Sam Harris asks, why do we keep trying to make the world theological? Asserting the existence of God doesn't really give us any more information than we had already. It is simply applying a label to the great unknown question of *why*.

The problem with this argument is similar to the familiar turtles problem we encountered in Chapter 7: we are labeling one turtle only to expose the turtle just beneath in. We're still left wondering where the world came from. In what realm does it reside? Who invested it with its rules? It's not clear that absence of God is any simpler an explanation for these mysteries than presence of one. Occam's razor may not prove the existence of God, but it doesn't seem to do much to disprove it.

A second defense of atheism is more indirect, drawing on the supposed consensus among scientists that God does not exist. Though most of the geniuses who created the Scientific Revolution were believers – Galileo, Newton, and Darwin among them – many of the most influential scientists of the last hundred years have been atheists, or at the very least agnostics. Carl Sagan, perhaps the most revered of the humanist scientists, asked, "How is it that hardly any major religion has looked at science and concluded, 'This is better than we thought! The universe is much bigger than our prophets said, grander, more subtle, more elegant?' Instead they say, 'No, no, no! My god is a little god, and I want him to stay that way.' A religion, old or new, that stressed the magnificence of the Universe as revealed by modern science might be able to draw forth reserves of reverence and awe hardly tapped by the conventional faiths." Richard Dawkins attempted to make a similar point when citing results from a survey of scientists about their theological beliefs. When asked if they believed in God, only seven percent said yes.

It turns out, though, that theological beliefs among scientists are far more nuanced. For example, in the survey Dawkins cited, to "believe in God" had a specific meaning: to "believe in a God that personally communicates with humanity." This definition ruled out belief in other ideas of God.

To many scientists, science itself illuminates the path towards God. "The first gulp from the glass of the natural sciences will turn you into an atheist, but at the bottom of the glass God is waiting for you," said Werner Heisenberg. Perhaps scientists haven't been willing to get to the bottom of the glass.

Einstein saw profound truth in the depths of the natural world: "What I see in Nature is a magnificent structure that we can comprehend only very imperfectly, and that must fill a thinking person with a feeling of humility. This is a genuinely religious feeling that has nothing to do with mysticism." Einstein is often mislabeled an atheist, but he rejected the term and in fact had a deeply spiritual outlook on life:

> [T]here is a third stage of religious experience which belongs to all of them, even though it is rarely found in a pure form: I shall call it cosmic religious feeling. It is very difficult to elucidate this feeling to anyone who is entirely without it, especially as there is no anthropomorphic conception of God corresponding to it. The individual feels the futility of human desires and aims and the sublimity and marvelous order which reveal themselves both in nature and in the world of thought ... The religious geniuses of all ages have been distinguished by this kind of religious feeling, which knows no dogma and no God conceived in man's image; so that there can be no church whose central teachings are based on it. Hence it is precisely among the heretics of every age that we find men who were filled with this highest kind of religious feeling and were in many

cases regarded by their contemporaries as atheists, sometimes also as saints.

The reality is that views among scientists vary widely; some are devout atheists, some are theists, and many, like Einstein, don't fall neatly into either camp. But to some degree it doesn't matter; there is a limit to how much we ought to divine about the ultimate nature of the world based on the philosophical beliefs of scientists. They are smart people, for sure, but they are humans just like us, and thus hold their own set of frames and biases. Even the best scientists are not necessarily credible beyond their area of expertise. A Ph.D. in the physical sciences does not guarantee insight into the ultimate nature of God and the Universe. We're better off reviewing the evidence we have and taking our own path towards the truth.

Contemporary atheists

For better or worse, contemporary atheism has been defined in the public imagination by its most visible, and often histrionic, advocates, particularly the Cambridge evolutionary biologist Richard Dawkins. He is famous both for the incisive clarity of his scientific explanations and his hubristic, often patronizing defense of atheism and criticism of religion in all forms.

Dawkins defends his worldview this way:

> If you're an atheist, you know, you believe, this is the only life you're going to get. It's a precious life. It's a beautiful life. It's something we should live to the full, to the end of our days. Where if you're religious and you believe in another life somehow, that means you don't live this life to the full because you think you're going to get another one. That's an awfully negative way to live a life. Being an atheist frees you up to live this life properly, happily and fully.

About God, Dawkins says: "Either he exists or doesn't. It is a scientific question: one day we may know the answer, and meanwhile we can say something pretty strong about the probability." He concludes that on a scale of 0 to 7, where 0 equals absolute proof that God does not exist and 7 equals complete faith in the existence of God, the evidence warrants a 1. "The God of the Old Testament is arguably the most unpleasant character in all of fiction," Dawkins writes. He is "jealous and proud of it; a petty, unjust, unforgiving control-freak; a vindictive, bloodthirsty ethnic cleanser; a misogynistic, homophobic, racist, infanticidal, genocidal, filicidal, pestilential, megalomaniacal, sadomasochistic, capriciously malevolent bully."

In person, apparently, Dawkins is amiable, mild-mannered and full of British courtesy. But in writing – and worse, on social media platforms like Twitter – he has an almost pathological need to bait, dogmatize, and oversimplify. About those who raise their children in a religious tradition, he tweets, "How dare you force your dopey unsubstantiated superstitions on innocent children too young to resist? How DARE you?"

Dawkins is particularly susceptible to ascribing malicious intent or trickery to those he doesn't agree with. In 2000, Freeman Dyson won the Templeton Prize, a somewhat controversial accolade among atheists because it "honors a living person who has made an exceptional contribution to affirming life's spiritual dimension, whether through insight, discovery, or practical works in his speech." In his acceptance speech, Dyson said, "I do not make any clear distinction between mind and God. God is what mind becomes when it has passed beyond the scale of our comprehension." Dawkins imagines another narrative going on inside Dyson's head, because Dyson, one of the pre-eminent physicists of his generation, can't possibly just *believe* this stuff. Surely he must have something else going on in his mind. Dawkins goes so far as to imagine what his internal narrative must be: "Have I said enough yet, and can I get back to doing physics now?"

Other atheists, like PZ Myers, strike an even more aggressive stance against theism. Myers drops the pretense of collegial discussion altogether, preferring instead a self-consciously provocative posture that occasionally strays into *ad hominem* attacks. He is notorious for a stunt in 2008, in which he mocked the Catholic rite of the Eucharist by treating the wafers, which Catholics view as sacred, "with profound disrespect and heinous cracker abuse." Though Meyers defended his behavior as satire, it represented a low point in the strained relationship between the atheism and religion.

Setting aside his public antics, Myers's argument for atheism is not particularly compelling. He offers the following story, a favorite of religious believers. A man goes to the doctor and learns he has cancer. He endures horrible treatments and misery, and is often at the edge of death. Eventually his cancer goes into remission and his prognosis improves. "Praise God for this miracle," he says. Myers points out that God is never held responsible for the cancer in the first place. A better story, he says, is that scientific medicine has saved him.

Myers's argument hinges on the assumption that the "baseline" we ought to expect, and are entitled to, is a life without cancer or a world without suffering. Perhaps, though, mere existence is a miracle in itself, as are rational humans who created scientific medicine. The gift of unexpected years of life can reasonably be seen as a miracle in itself.

This dogmatic opposition to anything that smells even vaguely of God-talk is unfortunate. It's laudable to be a free thinker and to defend the side of logic and reason when it's in peril. But in their zeal to "offend people's religion" at every opportunity, the most strident atheists expose their own limitations. They are so focused on disproving the traditional monotheistic conception of God that they don't realize they're trapped entirely within its frame. Religion has essentially defined and controlled the narrative of God in the West, and defined its very terms.

I'm reminded of the movie *Inception*, in which people can enter each others' dreams. Someone entering another's dream may be hostile, trying to kill the host, but they still must inhabit the subconscious world of the dreamer. Atheists are living in someone else's dream: the dream of monotheism.

The result is that our discussions about God wind up being constrained to a single dimension. On one end is the absolute certainty that God exists – theism. On the other is the absolute certainty God does not exist – atheism. A third choice, agnosticism, allows us to hedge between the Believer or Atheist camps, but we're usually still stuck on the same old number line. To use Dawkins's terminology, we may be a *de facto* theist (high probability that God exists but less than 100%); a theist leaner (somewhat higher than 50%); completely impartial (exactly 50%); an atheist leaner (less than 50%) or a *de facto* atheist (low probability but greater than 0%).

We need to break out of this stale debate altogether. If we inscribe a new path, hewing to reason and science but escaping from the monotheistic frame, where might it lead us?

God and mystical naturalism

To find God through a mystical naturalist framework, we must first make sure we're not working from an existing frame to prove or disprove someone else's prior conception. If we really want to find our own dream, we must avoid living in someone else's.

One way to do this is to reverse the usual approach to proving God's existence. Instead of positing a particular type of entity and then attempting (successfully or unsuccessfully) to establish a logical path to that entity, it makes more sense to observe what is, and then see if the label "God" seems to apply to that picture.

To me, it makes the most sense to define God as simply *everything* – the system of creation that includes all laws, matter, information, and anything else we can or can't conceive of. We are an inherent part of that system, and thus are a part of God. Or, as the philosopher Alan Watts observed, "Through our eyes, the universe is perceiving itself. Through our ears, the universe is listening to its harmonies. We are the witnesses through which the universe becomes conscious of its glory, of its magnificence."

You might contest that defining God simply as *what is* is tautological, just assigning a label to an already existing concept. Fair enough; as Paul Davies says, the existence of God is "a matter of taste." I might refine this a bit, and say that the *label* God is a matter of taste. We may use "the Universe," we may say the "magisterium," we might call it "nature" or "creation." It doesn't really matter. The ultimate nature of the world doesn't change based on the terms we use. We could call ourselves *ignostic,* the rabbi Sherwin Wine's word for one who believes it's pointless to discuss whether God exists, because the very term assumes too much about what God really is.

Whatever we call it, there is something profound about this scientifically consonant image of God. It is strikingly similar to the monotheistic, Abrahamic vision of God, with the signature distinction that it extricates the demon of duality. The image of God in mystical naturalism is all-powerful, in that it comprises all forces in the Universe. It is all-knowing, in that it comprises all information and processes. It is omnipresent: it is everywhere, at all times and beyond time. In its mystical order, it is ultimately wise. And while we can understand much of its inner workings, its very existence must remain an ineffable mystery.

Atheists ought also to embrace this definition because it posits nothing more than what is; it is the sum total of the natural world. It does not presuppose *how* the world came into existence. It does not require a humanoid male with a long white beard and a pointed finger acting at a distance, or any other received image of God. It requires nothing to be taken simply on faith. It adheres (as far as we can see) to the universal principles of order and sensibility. This comports well with contemporary views on atheism. Julian Baggini, in *Atheism: A Short Introduction*, writes: "What most atheists do believe is that although there is one kind of stuff in the Universe and it is physical, out of this stuff come minds, beauty, emotions, moral values – in short, the full gamut of human phenomena that give richness to human life."

This is what the nineteenth-century Hindu monk Swami Vivekananda was getting at when he distinguished the Western definition of atheism from the Hindu definition. In the West, atheism refers to people who do not believe in God. To Vivekananda, those who do not believe in themselves are atheists. All that's required to reject the label of "atheist" is to love your own soul.

In one fell swoop, the nondual *God That Is Everything* eliminates two millennia of philosophical baggage. Key questions that have dogged theologians are much easier to answer. For example, does God exist in or out of time? The passage of time, science tells us, is a story our brains create to help us make predictions that help us pass on our genes. God comprises time but also created it, by virtue of the ordered Universe that gave rise to human brains.

Similarly, we no longer need to ask whether God set the world in motion and then stepped back, or whether God takes an active role in our world. God *is* the motion of the world, the symphony of all its instruments working together.

Mystical naturalism implores us to see ourselves as an essential part of God. We ought to ask not "how and why was the Universe created?" but "Why did *I* create the Universe." Instead of asking "What is the nature of God?" – an ultimately unanswerable question – we might ask "What is the nature of awareness?" as awareness is, to borrow Alan Watts's phrasing, the eyes of God itself.

While the nondual conception of God seems to challenge our Western monotheistic notions, it fits more comfortably with Eastern spiritual traditions. In Hinduism and Buddhism, the idea that we are part of God is an essential part of the basic belief system. In Hindu thought, God is more naturally associated with the self. The ninth-century Hindu philosopher Shankara defined God as "that which permeates all, which nothing transcends and which, like the universal space around us, fills everything completely from within and without, that Supreme non-dual Brahman – that thou are."

In one illustrative story, a Westerner asks the guru Ramana Maharshi whether the great guru can help him see God. "I cannot help you see God," Maharshi answers, "because God is not the object. He is the subject, He is the seer, and you must concern yourself with finding out who the seer is. It must be inside you. You alone are God."

To assert that I am God is not as arrogant or as narcissistic as it may sound. In fact, it is radically egalitarian. Every person, every node of the Universe, has an equal claim. A child at a Syrian refugee camp may say, "How amazing am I that I created the moon and the stars." A tarantula and an asteroid, if they could self-reflect, might express the same thing. As the science writer Nancy Ellen Abrams poetically observes, God is something that emerges from human experience, "radiating from every human on earth, every book, every building and artifact, every plant grown by us, every shared endeavor, every mountaintop or galaxy that we have imbued with mythic significance."

The problem of evil

Before we move on, I want to take on one hard-core challenge to the existence of God: the problem of evil. If God is all-powerful and infinitely wise, why would God choose to create a world in which there is so much evil and suffering?

In traditional theology, the attempt to answer the question of evil without jettisoning God altogether is called theodicy. Central to this line of thinking is the concept of "ultimate harmony." While we see evil and suffering at the human level, God, in his infinite perspective, sees the bigger picture. God's morality is different than ours because God can see and understand all, while we have only a limited human view.

Perhaps, suggest some theologians, God has allowed evil into the world to allow us to fully develop as humans. This idea goes back at least to the second century AD and the writings of the Greek cleric Bishop Irenaeus, a key figure in the development of early Christianity. In order to become spiritually mature, we must be able to display moral virtue and express faith, and that's only possible if evil exists in the world. St. Augustine, in the fourth century, acknowledged that there is evil in the Universe, but argued that overall the world is better than it could be without it. The modern Christian perspective is summed up by Thomas Keller: "Just because you can't see or imagine a good reason why God must allow something to happen doesn't mean there can't be one."

How far can the idea of ultimate harmony get us? Theodicy has had no greater test than the Holocaust, the archetypal modern example of human suffering. Eugene Borowitz, a leading rabbi and theologian in reform Judaism, said, "Any God who could permit the Holocaust, who could remain silent during it, who could 'hide His face' while it dragged on, was not worth believing in. There might well be a limit to how much we could understand about Him, but Auschwitz demanded an unreasonable suspension of understanding. In the face of such great evil, God, the good and the powerful, was too inexplicable, so men said 'God is dead.' "

Elie Wiesel lost his mother and younger sister in the Holocaust, and later went on to become a renowned author and academic, winning the Nobel Peace Prize in 1986. During the Holocaust, Wiesel accepted God's will without questioning. God alone had greater plans. "God is testing us," Wiesel thought. "If he punishes us relentlessly, it's a sign that he loves us all the more." Eventually, however, his belief that the Jews were part of some greater plan was severely tested. "Every day I was moving a little further away from the God of my childhood," he said. But he did not lose his faith in God – only the picture of God from youth. When people asked how he still believed in God, he would ask them, 'How can you believe in man?' After all, God did not send down Auschwitz from heaven. Human beings did it."

Victor Frankl, too, endured the Holocaust and became well known for the account of his spiritual journey, *Man's Search for Meaning*. Frankl lived in Germany in the 1930s and was imprisoned in Auschwitz for three years, living in a shed with so many prisoners that there was not even room for each of them to squat on the ground. Prisoners received one five-ounce slab of stale bread to eat every four days. After he was freed, Frankl learned almost his entire family had been murdered by the Nazis.

And yet, even in this worst imaginable experience, Frankl found that those prisoners who sought meaning were those that survived, even prospered. Frankl tells of the evening when, while resting on the floor of their hut after another day's work of exhausting slave labor, a prisoner rushed in and asked them to run out to the assembly grounds to see the sunset. "Standing outside we saw sinister clouds glowing in the west and the whole sky alive with clouds of ever-changing shapes and colors, from steel blue to blood red.... Then, after minutes of moving silence, one prisoner said to another, 'how beautiful the world *could* be.' " At another point, Frankl was working in a trench in the dreary gloom of winter, struggling to understand his "slow dying" and trying to converse silently with his wife. "I sensed my spirit piercing through the enveloping gloom. I felt it transcend that hopeless, meaningless world, and from somewhere I heard a victorious 'Yes' in answer to my question of the existence of an ultimate purpose."

Mystical naturalism and theodicy

The experiences of Wiesel and Frankl illustrate that a certain conception of God is possible even in the face of ultimate evil. Our thinking about evil actually is rooted in the idea of unfairness: why do some of us experience great suffering, often caused by others, while some do not? The level of human joy and suffering is distributed drastically differently among us all. Some of us suffer in concentration camps; some of us relax on yachts in the Mediterranean; most of us exist somewhere in between.

Mystical naturalism, through its rejection of duality, gives us some tools to reconcile the seeming unfairness of the world with a benevolent creator. Unfairness relies on an individual, atomistic idea of consciousness – the idea that we are all eternally separate beings. Perhaps, however, each individual human consciousness is really part of a universal consciousness. Ultimately, then, that cosmic mind will experience everything that every human has experienced – suffering and pleasure, evil and goodness. The question then becomes: Is this holistic, complete set of all experience good, bad, or somewhere in between? If you add all of that experience together, creating a cosmic tally sheet of joy and suffering, a grand accounting of meaning and misery, does it come out in the black? In other words, is it better that all humans lived, or that none of them did?

There is no way, of course, to quantify the answer to this question. But the answer seems clear: it is better to have a real universe imbued with awareness, built of life and death, of joy and suffering, of awe and horror, all swirled together in an inextricable oneness, than to have nothing at all.

That simple truth seems clearer than ever as I stand out here underneath the vastness of the Rocky Mountain sky, the silvery light of a rising moon illuminating the length of the Collegiate Range, its snowfields glowing even in June. Spinoza's words ring in my head: "I say that all things are in God and move in God."

How sad that we've gotten so far from this pure, intuitive vision of God, so simple and yet so powerful, so uniting and yet so personal. To figure out how that happened, we'll need to traverse one of the thorniest regions of our journey: religion.

CHAPTER 9

Religion and its Discontents

My religion is very simple. My religion is kindness.

— Dalai Lama XIV

The next morning, I stroll through the quiet streets of Leadville. At one intersection, I see a curious sign. It reads "Leadville Jewish Museum – One Block" with a red arrow pointing down a small street. The metro areas of Colorado's Front Range have their fair share of Jews now, but it's a bit surprising to see a Jewish museum in a remote mountain town of 2,500 souls.

The museum is housed in an old synagogue that still periodically hosts services. As I step past the glass cases of memorabilia, I learn that in the late 1800s, Leadville was home to 300 Jews, who supported several synagogues and owned a number of downtown businesses, following the mining boom of the 1870s.

Seeing all this Jewish history right here in my adopted home state of Colorado, it's impossible not to feel some sort of connection with my cultural history. Genetically, I'm about as Jewish as it's possible to get. A few years ago I sent away for a $99 genetic test and learned I'm estimated to be 96.8% Ashkenazi Jewish, with the remaining 3.2% classified as "broadly European." All eight of my great-grandparents immigrated to New York and Philadelphia from the old country around the turn of the century. They came on ships, embarking from Lithuania and the Ukraine and Russia.

Like many secular American Jews, my relationship with my religion is fraught. Every year I look forward to the main holidays of the Jewish calendar, each of which holds personal meaning for me. Passover means assembling far flung family members and a few honored friends to share a feast of matzo ball soup, gefilte fish (which only Jews can truly love), and my mother's famous flourless chocolate cake. In the liberal, modern retelling, Passover is about remembering the oppressed among us, and fighting modern plagues like racism and homophobia.

Whenever the Seder gets around to the Plagues, its great fun to dip my pinkie in a glass of absurdly sweet wine and splash dots of it on my china plate – or, if there are kids around, to toss rubber bugs and Styrofoam hailstones around the dining room. But actually confronting the intended meaning of the plagues is more uncomfortable. It always requires some metaphorical sleight-of-hand to align our modern values with the vengeful theocratic deity who assaulted poor farmers with lice and murdered their children because they believed in the wrong god.

The actual Passover story is not as progressive as our modern retelling would have it. It is a tale originally designed not only to honor the powerless and perseverant, but also to demonize another race, the Egyptians – who, merely by dint of having a different culture, following a different religion, and living in a different land, were eligible to have their firstborn children slaughtered. Deuteronomy states: "When though goest out to battel against thine enemies ... though shalt save alive nothing that breatheth: But though shalt utterly destroy them; namely, the Hittites, and the Amorites, the Canaanites, and the Perizzites, the Hivites, and the Jebusites, as the Lord thy God hath commanded thee."

The first centuries of Judaism, between 700 BCE and 100 BCE, were filled with Jewish attempts at ethnic cleansing. Jews were constantly at war with their neighbors, fighting Egyptians, Romans, and Assyrians. In fact, the very story of Jews being enslaved to build the pyramids is historically questionable. Nowhere does evidence of a Jewish stay in Egypt appear in the historical record. "In short," writes Biblical scholar S. David Sperling, "the traditions of servitude in

Egypt, the tales of the Israelites wandering in the desert, and the stories of the conquest of the promised land all appear to be fictitious."

This is, perhaps, a harsh assessment of my assigned religion. Obviously the history of the Jews is littered with real oppressors, from the czars to the Nazis. In spite of my misgivings, I'm proud to be a Jew, an identity I share with 20% of Nobel Prize laureates. I like to claim commonality with Jews throughout history, from my scientist friends Spinoza and Einstein to comedians like Jerry Seinfeld and Mel Brooks, the archetypal communicators of the modern Jewish experience. When Seinfeld makes a joke about his relatives, he's really talking about *my* crotchety relatives. Even the fact that I'm challenging the roots of Judaism is, in a sense, a reflection of my religion. Judaism places a high value on philosophical discourse and educational inquiry, so in a sense, it's Judaism's fault that Jews like me challenge its core precepts.

I have good company in my very Jewish skepticism about Judaism. It was Einstein, in a 1954 letter to the German philosopher Eric Gutkind, who wrote:

> The word god is for me nothing more than the expression and product of human weaknesses, the Bible a collection of honourable, but still primitive legends which are nevertheless pretty childish. No interpretation no matter how subtle can (for me) change this. ... For me the Jewish religion like all others is an incarnation of the most childish superstitions. And the Jewish people to whom I gladly belong and with whose mentality I have a deep affinity have no different quality for me than all other people. As far as my experience goes, they are no better than other human groups, although they are protected from the worst cancers by a lack of power. Otherwise I cannot see anything 'chosen' about them.

As Einstein observed, Judaism has been "protected from the worst cancers" not because it holds some divine specialness, but because of a lack of power. Judaism's very value is rooted in the struggle of the Jews over the years. This very central trait of modern Judaism was not always a core part of the Jewish identity. It was only after their repeated attempts to conquer other tribes failed that Judaism took on its less aggressive modern flavor. My Jewish ancestors turned inward, focusing on reading the Torah and away from animal sacrifice. Jews survived because of the adaptability of Judaism. An important part of the story of Judaism is the triumph of education and rationality over the more genocidal rhetoric of the Old Testament.

Judaism, alas, is a religion – and it thus suffers from the same ills as any religion. In fact, the extent to which modern Judaism reflects contemporary morality is due more to our modern advances in rationalism and empiricism than to the word of the Bible.

I draw a few lessons from my experience with Judaism. It's common to see Judaism – like any religion – as something unified and eternal, as if it emerged from whole cloth 2,500 years ago. But – also like any religion – it is in fact a dynamic set of ideas created by humans that has evolved over time to suit contemporary purposes. According to some archaeologists and historians, the Jews were never a singular, unified group that had to defeat its enemies in order to live in the land of Israel, but rather a set of traditions that evolved from earlier societies, such as the Canaanites. These traditions were codified in the Bible for political reasons: to unify different sects.

The real roots of my ancestors' religious beliefs go back even earlier. The ancient Jews adopted their new religion from components of Babylonian culture. That culture emerged from previous belief systems, most of which have since been lost to history. Ultimately these patterns of beliefs emerged from the deserts of Eastern Africa many millennia ago. In spite of my genetic heritage, the vast majority of my ancestors predated Judaism. My African ancestors are no less tied to me, and their ancient traditions no less mine. We don't usually think of a Kenyan bushman with short kinky black hair and brown skin as a Jewish forefather, but indeed he was. These human forefathers lived in Africa for 90,000 years before they started traipsing across the desert. It was "only" 10,000 years ago or so that they started off on their long and winding tour of the Middle East, before settling down around the area now known as Israel.

The traditional conception of Judaism seems frozen in time because it reflects the moment when the technology of writing allowed us to create a permanent and unchanging record of our beliefs. Even so, Judaism has changed dramatically since it was first encoded in writing 2,700 years ago. In *Sapiens*, Yuval Noah Harari writes, "The political, economic, and social practices of modern Jews ... owe far more to the empires under which they lived during the past two millennia than to the traditions of the ancient kingdom of Judea. If King David were to show up in an ultra-Orthodox synagogue in present-day Jerusalem, he would be utterly bewildered to find people dressed in East European clothes, speaking in a German dialect (Yiddish) and having endless arguments about the meaning of a Babylonian text (the Talmud). There were neither synagogues, volumes of Talmud, nor even Torah scrolls in ancient Judea."

The question of what, exactly, it means to be a Jew is fraught. "Who is a Jew?" has earned its own Wikipedia page, covering various religious, ethnic, legal, and social definitions. In 2015, Israel's Minister of Religious Affairs decreed that Reform Jews cannot be considered Jewish, further exacerbating the gulf between conservative Israeli Jews and liberal American Jews. Even among progressive American Jewry, the view persists that Jews are defined on their matrilineage, so if one's mother is not a Jew, one is not a Jew.

What does all of this have to do with science, with seeking, or with what it means to be a human in the first place? The most profound lesson I draw from the history of the Jewish religion is that our religious identities are not essential, reified categories, but labels whose meaning changes through time. At the core, what I really am –

and what we all are – is human. (And even that identity may be too restrictive, and idea we'll explore further in Chapter 13.)

Attempts to reconcile science and religion

Religion is created by humans, but that fact, on its own, does not necessarily imply that religion is bad. Art and science are also created by humans, and both are essential to our well-being as a species. So let's take a moment to evaluate the rationalist arguments attempting to reconcile science and religion.

Francis Collins is the of the Director of the National Institutes of Health and a leading proponent of the idea that religion and science are compatible. In *The Language of God*, he begins by challenging the premise that God and science must be mutually exclusive. I've covered some of this ground in Chapter 8, in large part agreeing that science does not necessarily rule out certain conceptions of God. Where Collins, a Christian believer, runs into trouble is when he suddenly leaps from the ineffable, infinite sense of God as a source of goodness and creation, to the monotheistic, male deity of the Christian religion.

Collins argues in favor of the Abrahamic God by challenging the view that altruism can have arisen as the result of natural Darwinian selection. How, for example, could a sense of wanting to help someone without anyone else knowing possibly contribute to passing on our genes? This argument is faulty because it's easy to see how Darwinian selection could generate altruistic behavior. To envision a simplistic scenario: imagine ten separate cavemen societies, each genetically isolated. I live in one of the societies and I develop a mutation that makes me unusually likely to help someone. This may have a neutral effect on my ability to reproduce. But my progeny will carry the gene, and they will each be likely to help people. This will have a beneficial effect on the tribe as a whole, and the tribe will be more likely to survive than another, similar tribe.

And while altruism is surely an essential human attribute, we have plenty of anti-social tendencies too: selfishness, violence, xenophobia. Morality is not, as Collins suggests, something inherent in us and God-given, but a story we create with the tools and observations at our disposal. In fact, moral structures dispensed by religion can run directly counter to our thoughtful moral intuitions.

BioLogos, Collins's attempt to create common ground between his Christian faith and scientific truth, leaves a bit more breathing room. When faced with a biblical story that is historically inaccurate, such as Noah's flood, Christians have three options. First, they can abandon their faith and accept the scientific findings – but Christians, by definition, reject this option. Second, they can deny the scientific evidence to maintain their interpretations of Scripture. This option, with its long and sordid history, is obviously untenable if one is to respect scientific truth at all. So Collins offers a third option: to reconsider interpretations of Scripture in light of the evidence from God's creation. This yields the idea of BioLogos:

> At BioLogos, we present the Evolutionary Creationism (EC) viewpoint on origins. Like all Christians, we fully affirm that God is the creator of all life—including human beings in his image. We fully affirm that the Bible is the inspired and authoritative word of God. We also accept the science of evolution as the best description for how God brought about the diversity of life on earth.

> But while we accept the scientific evidence for evolution, BioLogos emphatically rejects Evolutionism, the atheistic worldview that so often accompanies the acceptance of biological evolution in public discussion. Evolutionism is a kind of scientism, which holds that all of reality can in principle be explained by science. In contrast, BioLogos believes that science is limited to

explaining the natural world, and that supernatural
events like miracles are part of reality too.

BioLogos acknowledges that evolution and a Creator can coexist. But it retains the dualistic flavor that so often infects these
discussions. Mystical naturalism, instead, holds that no distinction
needs to be made between the natural and the supernatural; all are
part of the same whole.

The theologian Timothy Keller attempts to unite reason and
religion by arguing on behalf of faith. In *The Reason for God*, he argues
that all doubts are actually a set of alternate beliefs: "You cannot
doubt Belief A except from a position of faith in Belief B. For
example, if you doubt Christianity because 'There can't be just *one* true
religion', you must recognize that this statement is itself an act of faith.
No one can prove it empirically, and it is not a universal truth that
everyone accepts."

Unfortunately, this argument, too, is logically flawed. It blurs the
distinction between a *belief* – something that is accepted on faith – and
a reasoned *conclusion* based on evidence and reason. In fact, the
argument that there can't be *one* true religion is strong and
straightforward: Religions, by definition, claim facts that are
demonstrably false. Any one religion, therefore, cannot be a true
depiction of reality. Moreover, doubt in one particular belief does not
need to imply acceptance of some other belief; for example, doubt
that the Bible represents the literal word of God does not imply
acceptance of any other particular belief.

Keller asks: "Has science essentially disproved Christian beliefs?
Must we choose between thinking scientifically and belief in God?"
Note the slight of hand here: Keller has crowded several different
ideas into one proposition, as if they are the same. "Christian beliefs"
and "belief in God" are wholly different beasts, and one can reject the
former and embrace the latter. A Christian belief is a specific fact
claim, such as that Jesus rose from the grave three days after his

crucifixion. But belief in God may take on a wide range of meanings, some of which are perfectly consistent with empirical facts.

Collins and Keller levy some high-powered logical thought in their attempts to unite reason and religion, even if their conclusions are flawed. Weaker arguments, such as the evangelist Ray Comfort's "How to Know God Exists," fail to take the time even to understand contrary ideas. For example, Comfort contests Buddhism's central precept that everything is an illusion, using this example: If you were a Buddhist standing at the door of an airplane, would you believe that gravity is an illusion? By acknowledging the very realness of gravity, we are meant to see the inherent flaw in Buddhism. This, of course, is a willful misrepresentation of what Buddhist thought actually has to say about what is illusory about the world. A better example of the Buddhist concept of an illusion is the self: an entity created by our brains that obscures the underlying, and very real, reality of which we are a part.

Fundamentalists like Comfort dress up their half-baked arguments in the guise of critical thinking. This approach, sometimes called "scientific creationism," has had notoriously disastrous results, including in the courts. One 1981 law in Arkansas that mandated equal time for creationism was overturned by the U.S. District Court on the grounds that it favored a particular religious view and thus violated the constitutional separation of church and state. It also pronounced creationism bogus science. Quite a few people who testified against the Arkansas bill were religious leaders and theologians.

The problem with most attempts to "prove" religion is that they begin with the narrow, rather than the broad. They focus on proving detailed fact claims, such as the assertion that Jesus rose from the grave. (Why, the scientific creationists ask, was his body never found?) Rationalists often get tangled up in the minutiae of refuting these individual fact claims. This Whack-a-Mole approach means that while each individual claim may be refuted, the core issue that religion is not a credible source of factual information is never addressed. A better starting point is to first assert where the burden of proof lies. It ought to lie not on those trying to disprove religious claims, but on those trying to prove them. Once again we should adhere to the Sagan Standard: "Extraordinary claims require extraordinary evidence."

What, then, is an "extraordinary claim"? Theologians like Alvin Plantinga say the claim that God does *not* exist is extraordinary, because the most natural conclusion is that God *does* exist. Atheists counter that no God is necessary to explain the natural world, and thus theists have the onus upon them to prove its existence. Neither approach seems quite right to me; the mystical naturalist idea of God-as-everything is both true *and* extraordinary.

With religion, where to assign the burden of proof is more clear-cut. Religious claims are highly specific and often arbitrary-seeming assertions about the nature of the world. They assert that Moses climbed up to Mt. Sinai to receive stone tablets from on high, and that Shiva held the Ganges on his head in order to prevent the destruction of Earth. These legends have poetic and metaphorical power, but if they are taken literally – as religious fundamentalists insist – it takes little to build a compelling case against them.

Why religion exists

If we are to judge religion strictly on its ability to deliver factual truths, it is bound to fail. But religion is more than that. Religions can also be sets of cultural practices, rituals, shared understandings, and devices to build community and foster resilience. Before condemning religion altogether, we must take into consideration these aspects.

We might start by asking why religion exists in the first place. Religion seems an essential aspect of the human experience. Our species has been stunningly successful, so for all its flaws religion must confer some benefits. The evolutionary explanation suggests that religion is a product of natural selection that has allowed our species to flourish across the millennia.

In *The Faith Instinct*, the science journalist Nicholas Wade lays out this argument in detail. All religions are branches of the same tree.

They evolved from the same central trunk, which represents the traditions of nomads in East Africa between 50,000 and 15,000 years ago — traditions that predate the written word and were probably centered around the life cycle and concerns of humans at the time: eating, security and reproduction. As humans spread from East Africa to other regions of the globe, they took traditions with them, which adapted to fit new settings and challenges. Those new traditions, beliefs, and revelations that tended to result in success for the community reproduced themselves through natural selection, crowding out the old ways.

Through this lens, religion may be seen as a feature of humanity, not a bug. We might not be here without it. Perhaps any species that's achieved our level of self-awareness and social complexity would have some similar mechanism to share and enforce social rules and to build resilience in desperate times.

Religion has also been proposed as having an important role in economic development. One of the first such theories was proposed by German sociologist Max Weber around the turn of the nineteenth century. Weber, considered the founder of the field of sociology, argued that the Calvinist work ethic, which combined hard work and asceticism, encouraged people to accumulate wealth and hold on to it rather than spend it. Similarly, Jews' economic success owes much to their focus on building business ties within the community.

An even more provocative idea is that science itself may owe its existence to monotheism. In *The Eerie Silence*, an inquiry into the mystery of extraterrestrial life, Paul Davies suggests that science itself was borne from monotheistic religions' linear view of the world, in which a single understandable force rules the Cosmos. According to the Bible, the world began at a specific point in time. Other events, such as the return of the Messiah, will occur (or have occurred) at specific points along the timeline, forever changing the world's spiritual status. This is distinct from the cyclical perspective of Eastern religions. For example, Hindu theology is based on *Kalachakra*, the great cycle of time. For Hindus, death is not the end, but merely the gateway to the next cycle.

Religion and happiness

It's clear that religion has been instrumental in the development of our species. But what about its role here, in the twenty-first century? On balance, is religion helping us or hurting us? Many avowed atheists believe that religion is the source of much of the world's deepest evil. Richard Dawkins laments: "The human psyche has two great sicknesses: the urge to carry vendetta across generations, and the tendency to fasten group labels on people rather than see them as individuals. Abrahamic religion mixes explosively with (and gives strong sanction to) both. Only the willfully blind could fail to implicate the divisive force of religion in most, if not all, of the violent enmities in the world today."

Scientific research doesn't quite bear out Dawkins's claim that religion is an unfettered evil. Participation in religious communities seems to make people happier, though that relationship is nuanced. It seems to provide something our secular, atomistic, highly independent culture does not – it builds a sense of community and social ties, which are essential to a sense of happiness. One study by researchers at the University of Texas found that people who never attended religious services had nearly twice the risk of death compared with people who attended more than once a week. Religious attendance improved social ties and caused more healthy behaviors. Sociology professor Rodney Start estimates that religiously motivated charity provides benefits of $2.6 trillion per year to the American economy. He attributes to America's rich religious culture reduced crime rates, better health, and stronger marriages.

The correlations between religious participation and positive outcomes, however, may be due to other factors. A Pew Research Center study found that people who described themselves as actively religious tended to be happier than those who were inactive or unaffiliated. But these people are also healthier, which is known to lead to greater happiness. Another study found that 33% of those who went to church every week and had close friends at work were extremely satisfied with their lives, compared to only 19% of people who attended church but did not have strong social ties there.

It turns out that the religious share a separate characteristic, which is that they are more likely to be engaged in other social groups. Religious believers seem to have better lives not necessarily because of their beliefs *per se*, but because of the indirect benefits of religious participation, which include better health habits, improved coping skills, and better social support. And religious practice doesn't help all of us. Psychology researcher Daniel Mochon and colleagues found that fervent believers benefit from religious involvement, but those with weaker beliefs are actually less happy than atheists and agnostics. The happiest countries are secular ones like Norway and Sweden. There, it is easy for atheists to connect with their fellow non-believers. Religious participation can also have tremendously negative effects if one's beliefs or identity are not aligned with religious dogma. An Australian study found that lesbian, bisexual, and gay people who were exposed to even subtle religious anti-gay prejudice displayed higher levels of stress, shame, depression and anxiety.

The conclusion: it's not what you believe, but that you believe it with other people, that makes you happy. In fact, the term "religion" itself comes from the Latin term "religio," meaning "bind together."

Religion has another benefit, one not captured by research studies. It creates a ready-made spiritual frame, giving participants permission and structure to perceive their lives more deeply than they might otherwise be willing to do. Religion invests the lives of its adherents with a sense of deeper purpose and meaning. To believers, God provides a sense that life itself is a miracle and that our lives have a meaningful plan.

I'm sometimes envious of this structure. If you're, say, a practicing Christian, the truth of God is invested in every aspect of your life. To be religious is to have permission to be spiritually intimate. It offers the idea of an essential "rightness" with the world, the idea that we are a part of some greater order, and that our job as humans is to accept and become part of that order. This idea is exemplified by the Serenity Prayer, which was written by the theologian Reinhold Niebuhr in the 1930s: *O God, give us the serenity to accept what cannot be changed, the courage to change what can be changed, and the wisdom to know the one from the other.*

Rationalists and atheists have lost religion, to our detriment. With all the dogmatic religious bathwater, we've thrown away the baby – a willingness to explore the spiritual possibilities of the natural world. Now we need to find it again, on our own terms.

How to evaluate religion

Religion is neither all "good," as Pope Francis would have it, nor all "bad," as Richard Dawkins believes. The truth is somewhere in between, in those shades of gray that are so hard for our brains to grab ahold of. Religions themselves vary greatly in their moral and ethical teachings, and in how well they represent reality. The natural, and combustible, question that comes next is: are some religions better than others? Each religion is different, so shouldn't we be able to use our rationalist tools to evaluate the pros and cons of each, and then, if we wish, to compare them? In twenty-first century America, this question is particularly fraught, as it often reduces to a referendum on the religion of Islam particularly, which has been accused of violence, intolerance, and misogyny.

I think it's worthwhile to address the question of whether some religious teachings and practices are simply better than others, as determined by a rationalist ethical framework. I'm not so interested in adjudicating among specific religions; I'll leave that task to others. A more useful starting point, I think, is to identify the criteria by which we should evaluate any religions. We ought to ask questions like these:

- To what degree do the decrees of the religion align with scientific fact?
- To what degree does the religion promote the idea that all people have fundamental equality?

- Does the religion suppress, tolerate, or encourage free thought and rational inquiry?
- To what degree do adherents of the religion update their beliefs in line with new scientific truths?
- To what degree does the religion encourage the idea that it is superior to others, and its adherents are superior to those of other religions?
- What is the quality of life of people practicing the religion, when other factors (like geography or resources or history) are taken into account?

Some philosophers have attempted to sidestep these questions by looking at religions' commonalities, rather than their differences. The Perennial philosophy holds that each of the world's religious traditions share a single, metaphysical truth or origin. Its origins date to the fifteenth century, when neo-Platonists like the Italian Renaissance philosopher Giovanni Pico della Mirandola suggested that truth could be found in many traditions, from Plato and Aristotle to the Quran and the Kabballah.

While it's true that each of these traditions have some elements of truth, they also each contain falsehoods. The Perennial philosophy doesn't tell us how to adjudicate among these beliefs. A more promising modern take on the idea is Aldous Huxley's, which has a less religious flavor: "The Perennial Philosophy is expressed most succinctly in the Sanskrit formula, *tat tvam asi*: the Atman, or immanent eternal Self, is one with Brahman, the Absolute Principle of all existence; and the last end of every human being, is to discover the fact for himself, to find out who he really is."

Buddhism

On our tour through organized religion, we've mostly focused on Western monotheistic religious traditions. We'll end the chapter with a different body of thought, a religion that may come closer to the tenets of mystical naturalism than any other organized spiritual system, probably because it's not a religion at all: Buddhism.

Siddhartha Gautama, later known as the Buddha, or the Awakened One, was born in about 567 BCE in Lumbini, near the foothills of the Himalayas in modern-day Nepal. He grew up in princely splendor in the palace of his father, a chief of the *Shakya* clan. But, the story goes, when he finally ventured beyond the walls of the palace, he was struck by the images of suffering he'd never seen before: a sick man, an old man, a corpse. He wandered in the forest for years, practicing a severely austere lifestyle, meditating, and supposedly eating a single grain of rice a day. Eventually he headed into a village in search of food and became enlightened, suddenly realizing there was nothing to gain because nothing had ever been lost. Out of this profound experience emerged the Buddha's insight into the nature of humanity, codified as the Four Noble Truths. All life is characterized by *dukkha*, or suffering. The origin of suffering is desire, a craving or thirsting for pleasure, or at a deeper level, for existence itself. The way to eliminate this desire is by liberating oneself from attachment. Finally, we may end suffering by following the eightfold path, or Middle Way, a prescription for living and acting rightly.

In some ways, Buddhism is more amenable to rationality than any of the other major religions. Desire creates suffering; liberation from suffering can be achieved by training the mind to experience reality as it is. Nobel Prize–winning philosopher Bertrand Russell described Buddhism this way:

> Buddhism is a combination of both speculative and scientific philosophy. It advocates the scientific method and pursues that to a finality that may be called Rationalistic. In it are to be found answers to such questions of interest as: 'What is mind and matter? Of them, which is of greater importance? Is the Universe moving towards a goal? What is man's

position? Is there living that is noble?' It takes up where science cannot lead because of the limitations of the latter's instruments. Its conquests are those of the mind.

The Dalai Lama, in a controversial speech before the Society for Neuroscience in 2005, identified the profound connections he saw between Buddhism and science in general:

> Although the Buddhist contemplative tradition and modern science have evolved from different historical, intellectual and cultural roots, I believe that at heart they share significant commonalities, especially in their basic philosophical outlook and methodology. On the philosophical level, both Buddhism and modern science share a deep suspicion of any notion of absolutes, whether conceptualized as a transcendent being, as an eternal, unchanging principle such as soul, or as a fundamental substratum of reality. Both Buddhism and science prefer to account for the evolution and emergence of the cosmos and life in terms of the complex interrelations of the natural laws of cause and effect. From the methodological perspective, both traditions emphasize the role of empiricism. For example, in the Buddhist investigative tradition, between the three recognized sources of knowledge - experience, reason and testimony - it is the evidence of the experience that takes precedence, with reason coming second and testimony last.

But the Dalai Lama goes further than just identifying these commonalities: he makes the quite radical assertion (for a religious leader) that, when in conflict, scientific findings should outweigh religious tradition:

In the Buddhist investigation of reality, at least in principle, empirical evidence should triumph over scriptural authority, no matter how deeply venerated a scripture may be. Even in the case of knowledge derived through reason or inference, its validity must derive ultimately from some observed facts of experience. Because of this methodological standpoint, I have often remarked to my Buddhist colleagues that the empirically verified insights of modern cosmology and astronomy must compel us now to modify, or in some cases reject, many aspects of traditional cosmology as found in ancient Buddhist texts.

As the Dalai Lama suggests, traditional Buddhism is not a fully coherent picture of reality. Like any pre-packaged belief system, it has its own convoluted lists of prescriptions and proscriptions. In addition to the eight stages of the Middle Way, there are three roots of evil (greed, ignorance, and hatred), five aggregates (material form, feelings, perception, volition, and sensory consciousness), and five precepts (abstain from killing, stealing, sexual misconduct, lying, and drugs). Sexual misconduct alone can refer to any of twenty different categories of forbidden liaisons, including those with parents, siblings, and "concubines who have been acquired by the ceremony which consists of dipping their hands into water."

All of this is simply a reminder that no religion – even Buddhism – is a perfect conveyor of the truth. That's why we need to ground our deepest spiritual understandings in science and reason. As the Dalai Lama observed:

All the world's major religions, with their emphasis on love, compassion, patience, tolerance, and forgiveness can and do promote inner values. But the reality of the world today is that grounding ethics in religion is no longer adequate. This is why I am increasingly convinced that the time has come to find a way of thinking about spirituality and ethics beyond religion altogether.

Religion has its uses, but I've argued in this chapter that those benefits can generally be better achieved through other means. Figuring out the truth of the natural world is better achieved through science and reason. The social and psychological benefits conferred by religion are a result of community-building, not faith, and so can be achieved through secular means. And an appreciation of the majesty and mystery of reality does not require religious belief; only attention to the wonder of the natural world in front of us.

There is one other important function of religion that deserves a bit more examination, and that's morality. How, without religion, are we to create the shared moral understandings that make a just and happy world possible? That's just the question we'll tackle next.

CHAPTER 10

The Invention of Morality

Vice, Virtue. It's best not to be too moral. You cheat
yourself out of too much *life*. Aim above morality. If
you apply that to life, then you're bound to live life
fully.

— Maude Chardin, *Harold and Maude*

For most of human history, morality was prescribed by
society. Our moral codes grew out of our belief systems –
first, animist folk traditions, and later, monotheistic religions
in which morality was received from God himself, eternal and fixed,
passed down through scripture or revelation. In a sense, this made
morality easy: the rules were all written down, translated from God's
own tongue, and humans' role was just to follow them. In the three
Biblical religions, the heart of moral teachings are the Ten
Commandments. The work of humans is not to debate the basic
premises themselves, but to properly apply these God-given
commandments to real life. To a devout Jew or Muslim, coveting thy
neighbor's wife is an indisputable moral wrong; the question only
comes in determining what counts as "coveting" and in assessing the
appropriate punishment.

The intuitive sense of morality we all hold owes its existence to
evolution. As our species developed, shared rules preserved our own
safety and protected us against threats. Ironically, this evolutionary
explanation may itself illuminate the importance of religion in
developing a society-wide sense of morality. In *The Faith Instinct,*

science journalist Nicholas Wade argues that while atheists are just as moral (or immoral) as the religious, they are moral because they are abiding by their community's moral standards. These standards, in turn, were shaped and enforced by religion. He proposes that the moral standards created through religion create a sort of "herd immunity" to bad behavior, much like vaccinations create herd immunity to disease, even though individuals may not be vaccinated and would be susceptible in the absence of the herd's protection.

The idea that morality is something that we create, rather than something handed down from on high, is anathema to many religious believers. Without an objective, external moral force, they argue, our moral rules are bound to be arbitrary or self-serving. But in actuality, it's religiously driven morality which is self-serving. Religion defines morality in circular terms, based on its own definitions. How do we know that our religion's moral teachings are correct? Because they comport with our moral intuitions. How do we know that those moral intuitions are correct? Because they comport with our religion's moral teachings.

Even modern religious believers would have a hard time taking literally the moral assertions of the Bible. Take this one from Deuteronomy: if your brother or son should entice you to worship other gods, "Thou shalt surely kill him; thine hand shall be first upon him to put him to death, and afterwards the hand of all the people. And thou shalt stone him with stones, that he die." These sorts of deific abominations make it clear that a literal interpretation of the Bible – or of any religious text, for that matter – is not necessarily a sure road to morality.

It is important to make a distinction between Biblical literalism and actual religious practice, which may or may not reflect this literal reading. Nicholas Wade argues, for example, that the type of ancient anecdotes that so rile up the New Atheists play no role in daily practice. It is the religious rituals that reinforce standards about what moral behavior should be. Jews see the Torah as the source of divine wisdom, but few synagogues I know of encourage stoning members of nearby churches to death, as is prescribed in Deuteronomy 13:10.

Sermons are more likely to be concerned with building community, supporting recent refugees in the community, or food.

Still, it's not a given that our modern religious practices are morally valid. But if religion is not to be used as the core source of morality, what *do* we use?

The answer, you'll not be surprised to hear, is reason. We've seen that rationality is a uniquely effective tool for plumbing the deeper truth of the Universe, so it follows that it is also the best tool for divining moral truths. This tradition goes back to Socrates, who believed we can only live a good life if we use reasoning to know what "good" and "evil" really are.

Morality is something we must create on our own. This is the premise of *constructivist morality*. (This idea should not be confused with its rightfully maligned counterpart, moral relativism, which says that morality differs depending on your perspective, identity, or cultural background.) Moral constructivism says that while morality is not delivered to us fully formed from on high, it is still a useful construct, and it ought to be a shared value. Morality is essential to run a functioning society, but only has meaning at the human level. We can't possibly conceptualize or execute a "universal" morality that somehow extends beyond our experience as humans.

Some would go so far as to say that even our intuitions about the most basic moral principles, such as human rights and fundamental equality, are merely ideas constructed by the social order. For example, the Constitution establishes an imagined idea that "all men are created equal," while the Code of Hammurabi describes an order with slaves, commoners, and nobility, each with different values. Which is right? To relativists like Yuval Noah Harari, neither are. Equality is created in the minds of all people, a shared myth on which we all agree.

I think this is a bit too pat. Using the tools of science and reason should tend to lead us closer to universal moral truths. As science has improved our understanding of human nature, so too have we come to appreciate similarities among different people. This has deep implications for moral equality. In the eighteenth century, phrenology – the allegedly scientific practice of measuring bumps on the head to measure mental traits – was used to justify the idea that whites were innately superior to other races. This conclusion had ethical implications; if blacks were mentally inferior, then a moral system which privileged whites was justified.

Decades of research in the fields of genetics, cognitive science, education, medicine, and sociology have convincingly shown that behavioral differences between racial groups are due to political, geographic, and cultural factors, rather than to innate differences. This conclusion, in turn, has pointed moral implications for our concept of equality.

Michael Shermer's virtuosic treatment of the topic, *The Moral Arc,* creates a compelling argument that the moral arc does indeed bend toward justice, and the primary cause for the bending of this arc is scientific rationalism: "We can trace the moral arc through science with data from many different lines of inquiry, all of which demonstrate that in general, as a species, we are becoming increasingly moral."

How far can we stretch rationality in service of morality? The school of thought known as "effective altruism" (EA) aims to find out. Popularized by the philosopher Peter Singer, this movement attempts to identify correct moral and ethical decisions through mathematics.

Let's say we're walking to work and see a toddler drowning in a lake. We can rush in to save him, but we may ruin our shoes and we might be late to work. Should we save the toddler? Most rational humans would say "yes." The inconvenience of ruining our shoes is much less consequential than saving a human life. Singer argues that we make that decision every day, because the cost to save a life in a developing country is only a few hundred dollars a day. (Subsequent research has revised this figure upwards; in 2015, the meta-charity Give Well estimated the cost to save a human life through its most effective charity, the Against Malaria Foundation, at $3,337).

The idea is that many moral choices are clear, once we've quantified them. This seems to work best when the metrics for success among two choices are similar. It's easy to decide between a charity that spends a million dollars to save a life and one that spends only a few thousand for the same result, given that the charities are otherwise similar. What's more difficult is to compare very different types of moral choices – such as the choice between donating to the Against Malaria Foundation or the choice to spend the same money for my town's homeless shelter. Even harder to assess are unquantifiable goods, such as support for the arts.

If we try to stretch our moral intuitions too far, we end up in a rabbit hole of unanswerable questions. For example, using an EA-type approach encourages us to quantify the moral value of creatures that can feel pain, including animals. But if we take this math to its extreme, we might calculate that it is better to kill all humans, because they have a negative net effect on animals. Similarly, if we create the most good by generating the greatest total happiness across all people, is it better to kill unhappy people? Or is simply the experience of being alive a net good, so we should create as many humans as possible? In that case, how do we incorporate rape into our moral calculations? A child born of forced intercourse is another conscious soul in the world, but that certainly can't mean that rape is a moral good.

The point is not that consequentialist morality is useless; quite the contrary. Effective altruism can make a huge improvement to our current ways of philanthropic giving. But once we stretch it to its logical philosophical extremes it breaks down. And that's because, as useful as morality is, it's ultimately a tool we've created.

The limitations of using intuition to determine morally correct actions is illustrated by the philosophical genre of "trolley problems," first formulated by British philosopher Philippa Foot in 1967. Imagine you see a runaway trolley barreling down the tracks. If it continues on its own track, it will certainly kill five innocent bystanders. However, you may move a lever that flips the switch, moving the trolley onto a side spur. If you do, the trolley will kill one innocent bystander. The question: what is the morally correct action?

Is it better that five people die, or is it better that only one person dies, even though you would be the cause of that death?

Innumerable permutations on the trolley problem have been constructed throughout the years. In the "Fat Man" version, you can save the five people by pushing a fat man over a bridge. This may be more morally objectionable because you must actively harm someone, unlike the original case, in which no harm was intended towards anyone. Another variant adds a spur which returns to the original track, so the death of the person on the spur prevents the certain death of the five originals. Still another changes the fat man to a fat villain who deliberately put the five people in peril in the first case.

The correct answer to all of these questions is that there is no correct answer. The whole point is to force us to examine our moralistic intuitions, and to find their limitations.

Another objection to using rationality to derive morality is known as the *is-ought problem*, first formulated by David Hume. *Is* statements are facts about the world as it is, but *ought* statements are moral or ethical imperatives. The problem, it's claimed, is that it's impossible to draw a logical *ought* conclusion from *is* premises. In other words, we cannot logically conclude a moral truth from statements about what is.

In *A Treatise on Human Nature,* Hume muses on how easily people slip from *is* to *ought*: "For as this *ought*, or *ought not*, expresses some new relation or affirmation, 'tis necessary that it should be observed and explained; and at the same time that a reason should be given, for what seems altogether inconceivable, how this new relation can be a deduction from others, which are entirely different from it." Moreover, he continues, "the distinction of vice and virtue is not founded merely on the relations of objects, nor is perceived by reason." Hume not only believed that reason can't get us to morality, he thought that there was no such thing as objective goodness at all: "'Tis not contrary to reason to prefer the destruction of the world to the scratching of my finger."

Is it true that morality cannot be determined by reason, as Hume claims?

1. I promised John I would pay back his twenty dollars.
2. A promise is equivalent to a commitment to pay John.
3. All else being equal, we ought to make good on our commitments.
4. In order to make good on my commitment to John, I ought to pay him.

There is a logical flaw in this type of reasoning, says Hume. It does not necessarily follow that the right thing to do in all situations is to honor our commitments – so Premise 3 is invalid. For example, I might be a heroin addict who scored my drugs from a dealer with the commitment that I pay him back tomorrow. The only way I can raise that kind of money in 24 hours is to pawn my grandmother's beloved wedding ring. Although I clearly have made a commitment to pay back the drug dealer, Hume would say, it's not clear that I *ought* to pay him back.

The problem is that the *only* grounds we have to determine morality is reason. Otherwise the very concept is empty. Morality without rationality is arbitrary. The only grounds we have to say we "ought" to do something is through reason – so we must either accept Premise 3 as logically valid (or invalid, depending on our reasoning process) – or else reject the idea of morality altogether.

We must strive to make moral rules objective, not relative. With perfect reasoning and perfect information about the world, two different people ought to come to very similar judgements about what is morally correct. Of course, our reasoning is never perfect, and we can never have all the information we'd like to make perfect moral judgements. We can only act morally or immorally relative to our knowledge.

This in itself raises some interesting questions. Our moral judgements are necessarily limited by the information we have access to and by our skills in critically evaluating this information. Is it immoral to make decisions with negative consequences if we make those decisions with the belief that they are morally correct? For example, suppose I am taught that the Bible is the inerrant word of God, and I lack the experience or critical thinking skills to conclude otherwise. I interpret the content of the Bible to mean that homosexuality is a sin. My teenage son expresses a romantic interest in boys, and in order to take the morally correct path, I send him to a religious camp that claims to "cure" my child of homosexuality.

In one sense, I acted correctly, because I followed my own moral sense. However, my actions are immoral when considered against actual facts of the world, which tells us that the Bible is not, in fact, the inerrant word of God, and that gay conversion camps are harmful and fail to change innate sexual preferences.

Perhaps it is useful, then, to think of two types of morality. First, there is our own personal responsibility to act morally *according to our own knowledge and understandings*. And second, there is objective morality – a Platonic point of moral perfection that represents those actions that best make the world a better place. We ought to be personally judged based on how well we realize the first type of morality, but as a society we ought to strive toward the second type of morality.

Even that Platonic ideal, though, doesn't really exist. Consider the case of consuming animals. I have personally concluded, through rational thought, that consuming an animal with a low capacity for cognitive awareness, like a chicken, is morally permissible, based on the information I have about the experience of being a chicken. But strong rationality arguments may be made forbidding eating chickens. There is no "right" answer to this question, at least one accessible to humans. Once again we have reached the bounds of morality, constructed as it is by human brains for human purposes.

Morality and free will

Morality seems to be closely tied to the idea of free will. In the dualistic framing, it is impossible to act morally if our behavior is determined, because we would have no control over it, and couldn't change our actions even if we wanted to. The theological answer to this conundrum is to posit a higher moral law which is somehow beyond the reach of the natural world. For example, C.S. Lewis offers a definition that distinguishes between laws of nature, which tell us what actually occurs, and moral laws, which tell us what *ought* to occur. We cannot help but obey the laws of nature, but we have a choice about obeying the moral law.

The philosophical term for this idea is known as the *Principle of Alternative Possibilities (PAP)*, formulated the libertarian philosopher Robert Kane. The PAP says that we are only morally responsible for what we do if we can do otherwise. Therefore, morality only comes into play when an action is not causally determined.

Contemporary compatibilists argue that morality is still possible even with determinism, imagining situations in which someone may be morally responsible but could not have done otherwise. Harry Frankfurt offers an example like this: imagine a bank robber, Donald, and his accomplice, Mike. Donald decides he's going to rob a bank and give half the proceeds to Mike, who's bankrolled him. Donald swears to go through with the robbery unless he gets a phone call from his daughter – let's call her Ivanka – as he's walking into the bank, in which case parental guilt will prevent him from carrying out the crime. On Robbery Day, Mike secretly switches Donald's phone to airplane mode, making it impossible for him to receive a phone call.

Ivanka is busy on a shopping trip to Moscow during the robbery, and so never makes the call. Donald robs the bank. He has thus committed an immoral act, even though there were no alternative possibilities.

I'm not sure, though, that this really gets us anywhere. Frankfurt's challenge to the PAP is really based on semantic interpretations of the term "morality." Is morality the act of choosing an ethically better behavior? Or is it some feature of the things we do, whether we can choose to do them or not?

Within the frame of mystical naturalism, the whole problem dissolves. Morality is both something we choose and a feature of those things, because "we" are synonymous with the laws of nature that determine events. When "we" act morally, what is really happening is that the laws of nature are acting in such a way as to create better outcomes for others.

This takes us back to the fundamental, mystical property of the Universe we explored back in Chapter 2. The Universe is ordered, and order inevitably gives rise to morality. Order is what allows evolution to occur, which in turn caused our brains to encourage us to behave altruistically. Order is what invests rationality with its power, and that rationality inevitably leads, over time, to a more moral society.

We cannot expect to ever reach the Platonic point of pure morality. Such a point doesn't exist. The best we can do is to use reason develop our moral intuitions as far as they'll go. And when that doesn't suffice, aim above morality.

CHAPTER 11

The Illusion of Time

The timeless in you is aware of life's timelessness.
And knows that yesterday is but today's memory and
tomorrow is today's dream.

— Khalil Gibran, The Prophet

As the summer burns on, my runs get longer. Today will be the longest training run of all: a 32-mile, eight-hour circuit through Rocky Mountain National Park. I start just before dawn at the Longs Peak Trailhead, and in an hour or so I've cleared treeline and the low sun has illuminated a field of boulders with amber light. I see the stunning rock face of the Diamond far ahead, and above it the soaring summit of Longs Peak. Today's circuit will take me up the side of the mountain, then down to the crystalline lakes surrounding Bear Lake and into the most touristed area of the park; then finally over Storm Pass and back to the trailhead.

A run this long can mess with your mind. Runners say they often end up running each mile three times: once before we've run it, worrying about it in advance; once while we're actually running it; and once after we've run it, evaluating and judging our performance.

My goal today, as I crest a ridge and the vast expanse of the entire park comes into view, is to run each mile only once. I am right here, right now, and I'll never pass this way again, at exactly this moment, in this way, in this place. In fact, this exact Andrew may never exist again. There is no past, no future, no now. There just *is*.

The nature of time

The true nature of time is one of the real blockbuster surprises science has revealed about the world. Time is not an essential, unidirectional component of ultimate reality. Rather, it's an emergent property of the Universe.

Time emerged shortly after the Big Bang. We cannot conceptualize what occurred "before" the Big Bang. As far as we know, there simply was no time before the Big Bang; it's like asking what is north of the North Pole. An alternative theory suggests that the Big Bang was merely the start of a new phase, and time stretches long before it – perhaps infinitely far into the past.

The classic scientific model of the world viewed time as a constant and immutable substance. Newton thought that time has its own, independent essence. In his *Philosophiae Naturalis Principia Mathematica,* Isaac Newton says, "Absolute, true, and mathematical time, of itself, and from its own nature, flows equably without relation to anything external."

All of that changed with Einstein's theory of relativity and the subsequent quantum revolution, both of which overturned the nature of time as constant and immutable. Einstein's theory of special relativity holds that under certain circumstances, the order of events is different depending on the standpoint of the observer. For example, consider two events on opposite sides of the galaxy: say, a supernova and the collision of a comet with a planet. From the perspective of an observer on Earth, the supernova occurs before the collision. But for observers elsewhere in the galaxy, the collision may appear to occur first. If we ask which precedes which in actuality, relativity says that there *is* no answer, because there is no absolute reference frame. Time is relative.

Imagine drawing a straight line in chalk on the sidewalk. Now draw two dots at different points somewhere along the line. Neither of these points is "ahead" or "behind" the other, and there is no "now" on the line to determine which are in the past and which are in the future. All of the points are equally real. Similarly, all points in time are real, and just exist at different points in sequence. That's why

Einstein said, "people...who believe in physics, know that the distinction between past, present, and future is only a stubbornly persistent illusion."

This truth reveals yet another deeply held assumption that is actually a human creation: causality. The very idea of causation is a story we've created to explain what appears to us to be a logical pattern of events. When we say an action in the present can cause something to happen in the future, we are observing a psychological phenomenon, not a physical one. Instead of a current event *causing* a future event, it's better to think of the two events as being *associated*. The current state of the world, in conjunction with the laws of the Universe, is associated with the state of the world a moment from now. We can easily run this association backwards: the state of the world a moment from now, in combination with the laws of the Universe, is *associated* with the state of the world in the present. This association allows us to make statistical predictions about the relationship between the world at these two time points, such as the prediction that if I release a Ming vase from my hand *now*, it will shatter on the marble floor a moment from now.

In fact, the phenomenon of time flowing continuously and at a consistent pace is an illusion our minds create. For example, sound and light travel at different speeds, and therefore arrive at our sensory organs at different times. Moreover, our various sensory organs take different amounts of time to process input. Your auditory system takes less than 1 millisecond to process sound, while your visual system takes more than 20 milliseconds. The result is that sound and light signals appear in your awareness at the exact same instant only when the source is about 10 meters away. If the source is closer or further away than that, you brain must essentially stop "live streaming" the experience and process it instead in chunks. Your brain then recreates a seamless, continuous movie.

Why, then, do we *feel* that time passes? Why do we have a sense of the past and the future, and why do we remember the past and not the future? Most likely, our brains evolved the capacity to imagine the future as a survival mechanism. This ability allowed our ancestors to make predictions about essential needs such as where to find food. Those of us who were better able to make correct predictions were at an evolutionary advantage, and passed along our genes to the next generation of time-thinkers.

According to one theory, our brains developed the capacity for this type of mental time travel at least 1.6 million years ago. Stone tools used by hominids at that time have been found many miles from where they were made. Why else drag along a tool when you've just eaten, unless you're planning to use it in the future?

Our brains generate a panoply of emotions to keep us focused on the future: fear, hope, anxiety, worry. What we ought to do is retain our sense of the future when it is *useful*, and to discard it when it is not. Emotions that bring us to this illusory future, such as fear or hope, are useful when they allow us to make our lives or those of others better.

The evolutionary hypothesis tells us why we'd want to have a concept of the future, but it doesn't explain why the future feels different from the past. The answer has to do with entropy, the amount of disorder in a system. A highly ordered system – such as an arrangement of cells in a tree limb – has low entropy. Throw the limb on a campfire, however, and the molecules disperse, increasing entropy.

The amount of entropy in the Universe has gradually been increasing since the Big Bang. All the matter in the Universe began in the singularity, an infinitely small, infinitely dense point, with zero entropy. Over time, the Universe has been expanding, and atoms have been free to gradually go about their way and create less ordered systems. Eventually, if the "open universe" theory is correct, the Universe will expand into eternity, with each atom gradually getting further away from each other one until there are no patterns left, just a formless cloud of particles too far apart to interact with each other.

Because of entropy, there are simply more disordered states than ordered states, so it is more likely that the Universe will assume a disordered state than an ordered one. Imagine your cat knocks a Ming vase off the counter and onto your marble floor. The vase – a single, highly ordered state of atoms – may now assume one of millions (probably many more) of disordered states. All those atoms *could* find themselves back in the highly ordered state of the unbroken vase, but there are many times more likely configurations that reflect a shattered vase, and thus, disorder.

This principle, called the Past Hypothesis rule, is why we cannot remember the future. If we see the Ming vase shattered on the floor and a cat cowering in the corner, there is a single, overwhelmingly likely, scenario for what happened a moment before. It's simply the most probable state of the world a moment before we walked into the room. We convert that probabilistic intuition into a belief in the past and future.

The idea that time has no independent existence has floated around philosophical circles for millennia, even if it has only achieved scientific legitimacy in the past century. Aristotle argued in the fourth century BCE that time exists only when people are around to observe it. Parmenides, in his poem *On Nature*, wrote, "there is not, and never will be, any time other than that which is present, since the fate has chained it so that it is whole and immovable."

Theologians have debated for millennia where God stands with respect to time. The doctrine of *ex nihilo* says that God created the Universe out of nothing and stands outside the world. Theologians from St. Augustine to St. Thomas Aquinas have understood God as a being standing outside of time who brought the material universe, including time, into existence. St. Augustine also thought that God is out of time: neither past, present, or future. God just *is*:

> Nor dost Thou by time, precede time: else shouldest Thou not precede all times. But Thou precedest all things past, by the sublimity of an everpresent eternity; and surpassest all future because they are future, and when they come, they shall be past; but Though art the Same, and Thy years fail not. Thy

years neither come nor go; whereas ours both come and go, that they all may come. The years stand together, because they do stand; nor are departing thrust out by coming years, for they pass not away; but ours shall all be, but when they shall no more be. Thy years are one day; and Thy day is not daily, but To-day, seeing Thy To-day gives not place unto to-morrow, for neither doth it replace yesterday. Thy Today, is Eternity.

This all raises some thorny theological questions: if God is outside of time, how does he intervene in human affairs? Can we speak to God in real time? Or has he already checked out, having long ago paid his cosmic dues by setting the world in motion, and letting us deal with the results best we can? Alfred North Whitehead, the great British philosopher of the late nineteenth and early twentieth centuries, proposed a novel understanding of God's place in time that was radically different than the prevailing (and still dominant) Western notion of physical reality. Rather than being made up fundamentally of material objects, reality consists of processes. Everything is moving and evolving, rather than fixed.

In process theology, God is no longer a fixed point outside the world, but inherent in the dynamic, ever-changing nature of the world. Process theism rejects the dualistic idea of creation *ex nihilo* – a God separate from his creation. Instead, in a sense all creation, including human beings, are part of God. Whitehead said, "It is as true to say that God transcends the World, as that the World transcends God. It is as true to say that God creates the World, as that the World creates God." This process-oriented view of reality has some similarities with contemporary scientific accounts, like that of the physicist Carlo Rovelli. In *The Order of Time*, he describes the world not as a collection of physical objects, but as a set of events unfolding.

For my money, it's Buddhism among all the religious traditions that has most successfully addressed the psychological illusion of time, replacing it with a sense of the sacredness of the everpresent now. Zen Master Dogen said, "Time is not separate from you, and as you are present, time does not go away. As time is not marked by coming and going, the moment you climbed the mountains is the time-being right now. If time keeps coming and going, you are the time-being right now."

All of this has some pretty deep spiritual implications. If every moment exists always, it's of utmost importance to be as present in every moment as possible, as this moment will last for eternity.

Uncertainty and the future

Isn't the future, you'd be reasonable to ask, uncertain in a way that the past is not, regardless of whatever psychological tales our brains are ginning up? Take the weekly Powerball lottery drawing. Last week's result is fixed, but this week's seems fundamentally uncertain. We can calculate a specific probability that any given set of numbers will come up – it's 1 in 22.877 billion, to be precise – but we can't know with certainty *which* numbers will appear until the moment arrives. What gives?

Probability is actually a statement of how much information we have about the world. If I flip a coin, the outcome – heads or tails – is fully explainable by the state of the world before I flip the coin and the laws of nature. In reality, though, we lack full information about the state of the world, so to us a given outcome is not a certainty, but a probability. When we say the odds of flipping heads are 50%, we are really saying that *based on our own limited information,* there is an equal chance of flipping heads or tails.

But if the future already "exists", in some sense, then isn't it possible that information can travel from the future to the past? In other words, isn't it scientifically plausible that we can receive messages from the future? It's true that the laws of physics can be run "backwards" just as easily as they can be run "forwards". However, the increasing entropy of the Universe over time means that from our perspective, the information we have about the past is far greater than the information we have about the future.

What about supposed intuitions about what will happen in the future – i.e., fortune telling? As we will see in the next chapter, there is scant scientific evidence for precognition. I believe that when we talk of premonitions we're actually talking about *right now*. A premonition that, say, something bad will happen in the future is actually not about the future; it's about our fears and anxieties here in the present.

Once again, we see the human tendency to assert some supernatural power to understand an imaginary future, instead of seeing that magic resides in the real world, right here and right now. Still, our scientific presupposition that we can't receive information from the future is based more on observation than airtight logic. Perhaps someday we will find a way to read signals from future time points that holds up to scientific investigation. Or perhaps time itself is nested within some greater order that somehow escapes our scientific tools.

The future is written, but it is also alive with possibilities. That's because the essential nature of being a human is to have imperfect information about the future. From our spiritual perspective, all things are on the table, and we are the driver of our own little corner of the Universe.

For now, it's enough just to be here, on the crest of a mountain ridge, and let the past and future float away into the distance.

CHAPTER 12

Life After Death and Other Psychic Phenomena

To the well-organized mind, death is but the next great adventure.

— J.K. Rowling, Harry Potter and the Sorcerer's Stone

About twenty years ago, I travelled from with my mother from my home in Maine to my grandparents' apartment on Long Island. My grandfather, 93, was nearing the end of a long and gentle life. When we arrived at the small apartment my grandparents had lived in since I was born, my grandfather was drifting in and out of consciousness, he'd stopped eating, and he often failed to recognize me, his only grandson.

That evening, we shared memories of my grandfather as he slept in his chair in the living room, his face unmistakably losing the sheen of life, hour by hour. I woke up the next morning with the unmistakable sense that a cosmic change had occurred. I knew my grandfather had died. I walked into the living room and what had been a person was now something different. The sight was not shocking or even disturbing, but it was deeply profound. It was a realization that I'd been living with an illusion all these years: the illusion that the material stuff in the shape of my grandfather *was* my grandfather. Of course, it was not; it was just a vehicle to house my grandfather's consciousness and mediate its relationship with the physical world.

My grandfather's passing was blessedly peaceful. He died not of cancer or heart disease but just because his body decided it was time to go. Until the days before his death, he was ambulatory and communicative; he wrote us poems and told jokes.

As an American, this was the most intimate experience I've had with death, and I'm grateful for it. In Western culture, death is hidden and forbidden. Our culture has medicalized death to such an extent that we have lost touch with its deeper meaning and its spiritual power. Three-quarters of deaths among people 65 and over occur in hospitals – those antiseptic, exorbitantly expensive palaces of pathology. Our attitude towards death reflects the way we pathologize "abnormalities" in general. If you feel off, you ought to take a drug; if you're having a baby, you're having a medical emergency. If you're dying, you should be in a hospital.

Our culture's frames for death are generally dark and frightening, and few align with science. Many monotheistic religious traditions see death as a portal to judgement through which only some of us will pass. New-age spirituality conjures speculative fantasies like past-life regression and ghosts and packages them as truth. And existentialists say that the inevitably of death calls into question the meaning of life itself. In *Denial of Death,* Ernest Becker put it this way: "What does it mean to be a self-conscious animal? The idea is ludicrous, if it is not monstrous. It means to know that one is food for worms. This is the terror: to have emerged from nothing, to have a name, consciousness of self, deep inner feeling, an excruciating inner yearning for life and self-expression – and with all this yet to die."

Mystical naturalism shows us something very different. It reveals that our modern concept of death is based on another one of those stories our brains have created: that of the independent self, marching ceaselessly forward through time. We've already seen that science endorses neither of these ideas. Both our perception of an individual "self" and our sense of moving through time are creations of our brains. When we adopt a more scientifically valid frame to transform our understanding of the self and of time, our interpretation of death changes just as radically.

All the world's belief systems have attempted to come up with a coherent explanation for the meaning of death. In African and Chinese religions, immorality is achieved by remembrance of those who are still living. Christians believe in a hierarchical progression in which the deeds of human life determine where we are placed in the eternal afterlife, whether Heaven or Hell. Souls are seen as unitary things which seem to arrive at their post-human destinations intact. Judaism emphasizes deeds here on earth — *mitzvot* — and doesn't make many pronouncements either way on the afterlife.

Eastern religions tend to take a less dualistic view of death. Vedantic Hinduism tells us that after death we are liberated from the cycle of existence and are unified with the One, and our concept of ourselves as distinct from the One is recognized as an illusion. Rumi, the thirteenth-century Sufi mystic, saw death and decay as an inevitable part of the endless flow of life.

Cycles of death and rebirth are an integral pattern in the real world as well, from the life cycle of the smallest mite to the birth of stars, which are born from the death of gas giants. The earth may only be habitable to humans because, a billion years into its development, a rock one-third its size smashed into it catastrophically, casting out trillions of tons of rock that coalesced into the moon. Without the gravitational balancing provided by the moon, the earth may not have been stable enough to evolve life.

Or consider the sudden death of the dinosaurs. They were among the most successful living things ever, reigning for 165 million years — almost a thousand times longer than our own species, *homo sapiens*. Dinosaurs were the very model of a successful, stable life form, yet they perished instantly when a meteor blasted into the Caribbean Sea 65 million years ago, throwing debris into the world's atmosphere and setting off a cycle of ecological disaster that destroyed the food chain on which dinosaurs relied. Without that catastrophic event, the variety of life that gave rise to humans would not have developed.

A key to understanding death through a mystical naturalist framework is to unite scientific findings about consciousness and about time. We don't fully understand how consciousness works or what makes it happen. There is no reason, however, to assume that it operates within our conception of time. If consciousness is an essential part of the fabric of the Universe, any given moment may exist "forever." Every moment in time "always" exists, just as the equator always exists even if I am north or south of it.

If the idea that we are alive, and *then* we are dead, is an illusion constructed by our evolutionarily greedy brains in order to get us to procreate and to preserve our genes, can we somehow see through this illusion? After death my "I" will no longer exist. I cannot suffer in death, because I will not *be*. (This idea goes back at least as far as Epicurus in the third century BCE, who believed that because after death we will no longer be able to feel anything, it's foolish to let the fear of death cause us pain while we are still alive.) In fact, from my perspective my own death is meaningless, because death is an attribute of my existence. When my body ceases to live and it no longer generates my unique "I," there will be no *thing* on which to hang the attribute of death. Andrew simply does not exist at that point in space and time. To die is an action verb that can only be done by a subject; without a subject – me – there is no meaning to death. I can't die any more than I can purple or I can 27. Our death is something salient only to those around us, who *are* aware and can experience that we do not exist at the same time point they do.

Life after death

Where are we, then, at time points at which our body is not supporting awareness – that is, when, to observers, we are dead? One hypothesis, endorsed by many religions as well as New Age forms of spirituality, is that we enter other realms or other bodies, roughly as our current selves. One alleged "proof" of this phenomenon are near death experiences, or NDEs. Self-reports of NDEs are often very similar. They usually involve people in a medically perilous situation, such as being cared for in a hospital in a life-or-death situation and under deep anesthesia. People who experience NDEs often perceive themselves as if outside of their body, and report experiences that seem impossible to account through conventional, naturalistic explanations.

Eben Alexander is a neuroscientist and author of *Proof of Heaven*, an account of his own NDE. After his brain was attacked by a rare illness, Alexander fell into a coma for seven days. During this time, he reports, he "journeyed beyond this world and encountered an angelic being who guided him into the deepest realms of super-physical existence. There he met with, and spoke with, the Divine source of the Universe itself." He writes that he alternated between being in the ICU and experiencing that he was ready to skydive out of an airplane, passing between the tops of two enormous puffy white clouds. He travelled to the "Realm of the Earthworm's Eye View, the idyllic Gateway, and the awesome heavenly Core."

Alexander claims that his experience was not explainable by science: "Without a functioning neocortex," he says, "the limbic system could not produce visions with the clarity and logic I experienced." He dismisses other neuroscientific hypotheses – such as the idea that he had experienced REM intrusion, which could generate the same experiences – because they would require a functioning neocortex. Ultimately he concludes that he did, in fact, experience a literal, post-death, realm, and the explanation lies (somehow) in the mystery of consciousness itself.

Kimberly Clark Sharp is another "true believer" in NDEs. In her book *After the Light*, she relates the story of a patient named Maria who was rushed to the hospital after a severe heart attack. Several days later, after going into cardiac arrest, Maria found herself floating outside the hospital and spotted a single tennis shoe on the ledge of the north side of the third floor of the building. After regaining consciousness, Maria was able to report not only the location of this shoe, but also precise details concerning its appearance, such as that its little toe area was worn and one of its laces was stuck underneath its heel. Upon hearing this bizarrely precise description, Clark found the shoe exactly where Maria had described it. Clark concluded, "The only way she could have had such a perspective was if she had been floating right outside and at very close range to the tennis shoe."

As scientific seekers, where do we begin to engage with accounts like these? We ought to frame our inquiry around a naturalist perspective, which assumes that the workings of the world, including the workings of our brain, must be consonant with reason and science.

We can start by asking whether a credible mechanism exists to explain NDEs. Are there brain functions or physical phenomena (quantum mechanics, say) that suggest a plausible mechanism for NDEs? In the example of Maria's shoe, do we have a theory of how the information about the shoe on the hospital window ledge might have been transmitted to Maria's mind? For Maria to absorb and then report this information, it somehow must have altered the physical state of her brain. Unfortunately, there are no credible explanations for how this might this have happened within the extremely convincing framework of naturalism.

There are theories that challenge this naturalistic framework, but they lack any sort of scientific credibility, and they are mired in dualism. They suppose that consciousness and the physical world somehow operate on separate planes, without any connection between them. The book *Irreducible Mind*, written by a team led by University of Virginia professor Edward F. Kelly, advances such a theory. Under the guise of academic flourishes, it throws together NDEs, mystical experiences, psi, and anything else that seems "magical" into a supposed argument for a dualistic mystical reality largely divorced from reason.

Unfortunately, none of this comports with our detailed factual understanding of the natural world. Accounts like Alexander's and Kelly's do not offer a coherent explanation to describe how information travels from the non-physical realm to the physical one. When Alexander experienced something and converted his experiences to words, certain molecules in his brain moved. What physical force caused them to move? Did they break the laws of physics? If so, his explanation would fall clearly into the realm of extraordinary claims, which in turn require extraordinary proof.

Proof of Heaven has been lucrative for Alexander, selling millions of copies in thirty-five countries. Its central claims, however, have been thoroughly debunked by both the medical staff that observed the operation and by neuroscientists, who have proposed other, more compelling explanations for his experiences. Sebastian Dieguez, a clinical neurologist, observed that *Proof of Heaven* is "painstakingly redundant, astoundingly arrogant in its claims and intents." The authors take reports of paranormal phenomena and wild claims at face value and trip right over the familiar stumbling blocks of quantum mysticism and other esoterica. They ignore, rather than build upon, the rich and valuable web of scientific understanding about the mind.

A better tack would be to explore known neurological mechanisms and see how well they fit Alexander's experience. Oliver Sacks concluded that "the one most plausible hypothesis in Dr. Alexander's case ... is that his NDE occurred not during his coma, but as he was surfacing from the coma and his cortex was returning to full function. It is curious that he does not allow this obvious and natural explanation, but instead insists on a supernatural one."

Mystical naturalism and life after death

We can rule out NDE's like Eben Alexander's as a compelling explanation for what happens after death. Perhaps surprisingly, though, mystical naturalism does not rule out life after death in general. In fact, I'd like to argue just the opposite: that scientific understanding is most consistent with a belief in life after death, just not the pseudoscientific or superstitious versions we're familiar with.

To say that again: science and reason can credibly establish the likely existence of life after death. But to build a strong and scientifically credible case, we'll need to make some semantic adjustments. Let's start by examining each of the three terms in the question: life, after, and death.

First, "life." In the usual framing of this question, "life" means something like "conscious awareness." When we ask whether there is life after death, we mean something like, *will my conscious awareness continue to exist after my physical body dies?*

The problem with this conception of "life" is that we associate one "conscious awareness" with one body. But as we've seen, the concept of the self is a story generated by the brain. When we die, the box that contains the conception of ourselves dissolves, like the moat of a sandcastle. We don't know what happens next, but it's plausible, though not certain, that our individual awareness rejoins the ocean of the greater whole, the world consciousness.

Next let's look next at that term "after." The sense of past and future is a psychological experience caused by increasing entropy in the Universe, but it shows up nowhere in the laws of physics themselves. So when we ask if something will happen *after* some other thing, we are applying a psychological construct to something as cosmically inexplicable as the experience of consciousness. There is no reason to think that consciousness itself must remain within the linear, unidirectional chronology we perceive.

Finally, the word "death." The conventional interpretation of the term seems more or less obvious. My grandfather passed away in 2001; he is *dead*. I am writing this right now; I am *alive*. We may quibble about the boundary situations, such as whether someone in a vegetative state is alive. But in general, "death" seems relatively easy to identify.

Again, though, there are many presuppositions built into this idea of death, and these come not from any rigorous scientific inquiry, but from our cultural and religious assumptions and traditions. Bodies can surely die, in that they are no longer able to perform the functions we associate with human life, such as metabolizing food or processing language. But "I" am not "my body." There is no conscious "me" that is connected in a one-to-one, inviolate relationship with certain molecules in the physical world.

One fascinating example that illuminates this principle is patients who have been in an accident and had a limb removed. Nerve systems that previously led to the removed limb can be rerouted to another point on the surface of the body — the stomach, for example. Now, when the stomach is touched, the subject perceives that their removed limb — a leg, say — is being touched. Those neural wires still connect to cognitive structures in the brain representing "leg."

The point is that our brains are not showing us the "actual" world, but are showing us a representation of it. The fact that we can "trick" the system shows that the connection between the physical world and our conscious awareness is a tenuous one. Our bodies can die, in that they no longer support consciousness, but that is altogether different from saying that "I" die.

If the awareness of which we are part is eternal and we are really part of a unified whole, then life after death makes more sense. Of course, while this is scientifically *plausible*, it's not a true scientific claim. We don't really know how consciousness works; we're just making educated guesses. What we can say with certainty is that we don't know what happens after our bodies die, but science does not rule out some sort of existence beyond our terrestrial, corporeal experience.

Psi

NDE's are part of the broader domain of parapsychology, or *psi*. Parapsychology is often thought of as the embarrassing red-headed stepchild of "real" science. In addition to NDEs, research into psi investigates alleged phenomena like psychokinesis (the ability to move items with our minds) and precognition (the ability to read the future).

For years, the Princeton Engineering Anomalies Research (PEAR) lab was the most prominent research organization investigating parapsychology. Dean Radin, former President of the Parapsychological Association, claims that its researchers "have produced persuasive, consistent, replicated evidence that mental intention is associated with the behavior of physical systems ... the experimental results are not likely due to chance, selective reporting, poor experimental design, only a few individuals, or only a few experimenters." Joseph Selbie, the popular author of books about the intersection of science and New Age spirituality, claims that the quality of science conducted by PEAR was of "overwhelmingly high quality," and that "hundreds of volunteers, in thousands of experiments, accumulating billions of data points" revealed that "nearly every volunteer had successfully altered the baseline distribution of the random event generator (REG) to an overwhelmingly statistically meaningful degree."

This conclusion, however, is belied by more rigorous examinations of PEAR's research methods. Robert T. Carroll, editor of Skepdic.com, wrote: "perhaps the most disconcerting thing about PEAR is the fact that suggestions by critics that should have been considered were routinely ignored. Physicist Bob Park reports, for example, that he suggested to PEAR founder Robert Jahn two types of experiments that would have bypassed the main criticisms aimed at PEAR. Why not do a double-blind experiment? asked Park. Have a second REG determine the task of the operator and do not let this determination be known to the one recording the results. This could have eliminated the charge of experimenter bias."

One of the main problems with PEAR's research is that its results seem to be selectively reported. This is the unfortunately common practice of "p-hacking," which occurs when an experiment's results are cherry-picked to find just the ones that are statistically significant. To avoid this, one must state ahead of time the specific statistical comparisons that will be conducted once results are collected. P-hacking is endemic throughout the psychological research field. Studies have found that academic publishing houses frequently publish results that falsely conclude an effect is present, when in fact there is no effect at all – what's known to research methodologists as a "Type I error." This problem is well-known in the research community; Stanford University professor John Ioannidis's landmark 2005 paper, "Why Most Published Research Findings Are False," has since become one of the most downloaded scientific articles of all time.

The real problem with PEAR was not that it was investigating a non-traditional area. It was that it failed to adhere to standards of high-quality research inquiry. True believers like Selbie intimate that academia conspired to hide PEAR's research findings: "No prestigious scientific journal ever published their papers. PEAR's findings, by every objective measure, were based on facts gathered in a stringently scientific manner. But because their findings fell outside scientific orthodoxy they were not given any scientific legitimacy."

This is the old academic conspiracy theory trope: the "real" results are too challenging and groundbreaking to ever be allowed past the institutional gatekeepers. This is the conspiracy theorist's last refuge, used to defend any number of false claims, from climate denial to voter fraud.

In fact, there is a huge incentive to publish credible research findings that challenge the conventional scientific wisdom. Time after time – from heliocentrism to relativity – experimental research has validated new theories that profoundly challenged the current paradigm. Anyone able to produce scientifically credible evidence of *psi* would instantly become a hero, and probably a millionaire. Any skepticism would easily be erased by replications of the same study, which would show similar results and vindicate the original researcher.

The path of scientific progress is imperfect and does not proceed in a straight line, but eventually, it is quite good at uncovering the truth. The pioneers who instigated scientific revolutions, like Galileo and Einstein, were not all honored in their time, but they ultimately became heroes.

The poster child for pseudoscientific claims about mystical phenomena is undoubtedly Deepak Chopra. He claims that "science has tons of data about phenomena that don't fit any explanation." As an example, he poses these three questions:

1. How does an observer cause light to change from acting like a wave to acting like a particle?
2. How can a group of ordinary people cause a random number generator to turn out more ones than zeros simply by wanting it to?
3. Where in the brain does the self live? Why do I feel like myself and no one else?

These questions are quite different from one another. The first indicates a clear misunderstanding of the term "observer" in quantum theory. While quantum mechanics is certainly mysterious, the collapse of a wave function does not require a human observer, nor does the act of observation in some way change the behavior of light.

The second question relies on a premise that is simply false. As we just saw, no credible studies have found any sort of parapsychological effects of the type Chopra claims.

The third question, though, is wholly different. This is a legitimate, deeply philosophical, and intensely mysterious question, one upon which science may help us shed some light, but which still remains largely beyond our grasp.

What we need is to distinguish misleading or pseudoscientific questions, like the first two, from true philosophical and spiritual questions, like the third. We do this not by eschewing science, but by embracing it.

Mysticism without pseudoscience

For his 2003 book *Rational Mysticism*, the science journalist John Horgan interviewed many mystics, most of whom were intelligent, responsible people, yet a surprising number believed in parapsychological phenomena like extrasensory perception and telekinesis. Why are otherwise sharp people so drawn to pseudoscientific explanations like NDEs?

I think the answer is rooted in our culture's dualistic separation between the physical world and the spiritual world. We assume implicitly that an "everyday" experience (i.e., one in which we understand the physical mechanism) cannot be a spiritually profound one. We feel that the deeply meaningful and mystical must exist beyond our realm. We are continually trying to drain the miracle out of the physical world right in front of us.

These miracles inhabit every moment of our lives, if we are able to take a moment to observe them. I can cause a pen to move and inscribe letters on a piece of paper sheerly with the power of my mind. I can will another person to act in a particular way by my mental powers alone. If I have the desire for another glass of wine at dinner, these intentions are somehow able to affect the physical behavior of my vocal cords, which create sound waves that travel to my server's ears. These waves are then magically converted by his brain into electrical impulses, and then to a motivation in his brain to pour me some more wine.

We can see this miraculous spirit world around us every day: when we look up at the night sky and see baby galaxies just beginning their own ordered march towards growing planets and perhaps life; when we examine the secrets of math; when we contemplate the true nature of time. If the natural world *is* the spiritual world, we don't need to look further to find miracles. We are already living in one.

CHAPTER 13

Being Human

I am what time, circumstance, history, have made of
me, certainly, but I am, also, much more than that.
So are we all.

— James Baldwin

I'm in the midst of a restless, broken sleep when the tinny alarm
on my running watch rings. It's 2:50 AM, and today I am
running the trail race that caps off my running season: the
Never Summer 100k. On its 61-mile journey through the remote
mountains of northern Colorado, I will climb and descend 13,000 feet
over three mountain peaks. In about an hour and a half I will start the
journey, and if I'm lucky I'll finish by midnight.

A few friends and I have rented a cabin near the race's start, in
the tiny town of Gould. I hop out of bed before anyone else is awake
and prepare for the day ahead. *Why am I doing this?* I ask myself.
Knowing your *why*, after all, is the most important piece of gear for
an ultramarathon; not the right shoes, food, or clothes. If you don't
have a *why*, you're liable to strand yourself when the going gets tough.
But if you do, you can make it through just about anything.

I think back to that quote by Aldous Huxley: "the Absolute
Principle of all existence; and the last end of every human being, is to
discover the fact for himself, to find out who he really is." That's why
I'm here today. To find out who I really am.

Who *am* I, from a philosophical perspective? If my identity as in an independent self is a construction of my brain, am I anything? Am I everything? To frame the question another way: What does it mean to be a human?

Our brains, and our culture, have programmed us to think of a "human" as an essential and indivisible core component of the Universe. Most religious traditions give great import to the spiritual primacy of humanity. Our social and moral structures acknowledge our relationships with non-human life forms – animals, for example – but they rarely challenge the notion that a human is a strictly-defined and supremely important category of life.

This is an example of *essentialism*, and it pervades our ways of thinking about ourselves. Our essences are those things whose existence seems so obvious they hardly need defending. Take gender: for most of the history of American culture, people have been identified as either male or female, and we've assumed that strict categorization is a prepackaged, *essential* part of what it means to be human. The Bible even details a story for how these two essential sides of humanity came to be: woman was created from one of man's ribs.

This division into two genders works, except when it does not. We might ask: What is it that categorizes people into one of these two genders? One answer is our chromosomes: humans are born with 46 chromosomes in 23 pairs, and the last pair, which is either X or Y, determines a person's sex. A human female's genome is designated as 46XX, while a human male is designated as 46XY.

But millions of people have some other combination. Some people are born with a single sex chromosome (so they are 45X or 45Y) and some have three or more sex chromosomes (so they are 47XXX or 47XXY). In addition, some males are born as 46XX. Even people with typical chromosomal makeups (46XX or 46YY) may experience developmental anomalies before they are born, resulting in anatomy which doesn't fit the typical definitions of female or male. For example, they may appear to be female on the outside, with a clitoris or labia, but have anatomy typical of males on the inside. Or a boy may be born with a scrotum divided so that it has formed more like labia. People who are biologically fully female or male may identify as the opposite gender, or they may not identify strictly as one of the two genders at all, preferring terms like "pansexual" or "gender queer."

Gender obviously has some useful meaning, and humans – and their mammalian forebears – evolved to predominantly feature two sexes. But "male" and "female" are simply terms we have invented to describe this general pattern. We might have turned out like clownfish, some of whom begin life as males and change to females. Or we might have developed like wrasses, a reef fish whose females may transition to males if another male leaves the group. Or banana slugs, which can fertilize themselves if they have trouble finding a date with another banana slug.

Racial essentialism is even more pernicious. For much of Western history, "white" and "black" were thought of as clearly distinct, essential groups with defined intellectual, physical, and even moral characteristics. Now we know, of course, that while race meaningfully describes some human characteristics – such as skin tone – it does not describe others, such as mathematical ability. Furthermore, race is continuous, not discrete. Each of us, in a sense, has our own "race," which describes the specific combination of geographic locations in which our ancestors evolved.

The African American scholar Henry Louis Gates famously discovered this when he underwent genetic testing to determine his ancestry. In the 2006 T.V. special *African American Lives*, he traced his genealogy, expecting to find primarily African DNA. Instead, he learned that he has 50% European and 50% African ancestry. Is Gates still black? Is he white? Is he something else altogether? Does it matter?

The idea that race and gender are fungible, human-created concepts is no longer controversial to most rationalist thinkers. They are creations of our mind, labels to help us understand – and often misunderstand – a pattern we observe.

What's less obvious, but even more important, is that the term "human" is a label we create to describe a pattern, too. The concept of a "human" is a fabrication, an invention that sometimes enlightens us as to the true nature of existence, and sometimes obscures it. What we call a "person" is a pattern that our brains have created and then filled with "stuff" – the stuff we experience as an indivisible Self. Our societal structures have wrapped a set of moral and religious beliefs around this pattern, making us forget that we created the construct in the first place. The Bible tells us God himself looks like a human; we were created us in His own image. Our legal and social creed tells us that human life is sacred in a way that is unique to our species.

It's impossible not to see the circularity once you look closely: our human brains created the idea of a "person," and then those "persons" created beliefs to back-explain our human brains. We created the idea of human exceptionalism to convince ourselves we truly are exceptional.

Even so, the idea that there is something special about being human forms the foundation of an egalitarian democracy ("…all men are created equal, with inalienable rights"). If taken seriously, it's a profound step forward from the tribal, hierarchical moral structure that came before it, in which people who were not of the correct gender and skin color were not considered fully human.

This is the driving force of humanism, an attempt to create a belief system around the idea that moral rules come not from God, but from humans ourselves. The American Humanist Association defines its worldview as "a progressive philosophy of life that, without theism or other supernatural beliefs, affirms our ability and responsibility to lead ethical lives of personal fulfillment that aspire to the greater good."

Humanism has been criticized as a belief system that actually reflects a worship of humans above all other creatures. In *Sapiens*, Yuval Noah Harari writes, "All humanists worship humanity, or more correctly, Homo Sapiens. Humanism is a belief that homo sapiens has a unique and sacred nature, which is fundamentally different from the nature of all other animals and of all other phenomena."

This doesn't quite capture the heart of humanism. There *is* something special about our species. The fact that we possess self-awareness and have the capacity for high-level thinking and reflection is surely meaningful. But that doesn't mean there aren't other forms of existence that *aren't* special. Harari also sees an inherent conflict between humanism, which glorifies humans, and the belief that the highest end is glorifying God, as the Protestant settlers believed. For mystical naturalists, this distinction vanishes. Humans are simply an aspect of God, and when we worship one we worship all.

What is human?

We think of a "human" as an easily definable category, but in fact it's not always so easy to say what is a human and what is not. This is nowhere clearer than at the beginning of life. The abortion debate is so intractable largely because it requires us to define the precise point at which an entity becomes "human." Does life begin at conception, as Catholics believe? Or at birth, as abortion rights supporters believe? Or somewhere in the middle?

The answer is that there is no right answer, because "human life" is a construct our brains have invented. The correct solution to the abortion problem is to use rationality to best evaluate a complex situation. We ought to ask questions like: what level of consciousness does a life-form attain at various stages in embryonic development? What moral weight should we give to life-forms with this level of consciousness, and why? How would the world be better or worse in the absence or presence of this life-form when it matures? These are all complex questions with no one right answer, and our political system is not well suited to address them. As much as possible, we ought to adjust our political system so it is able to accommodate rationally-based, complex arguments.

The fuzziness of the definition of a human would have been more obvious several hundred thousand years ago, when we shared the earth with another form of humans, the Neanderthals. Our *homo sapien* ancestors coexisted and even mated with them. At the time, our species was not the unchallenged, self-aware master of life on Earth, but just another one of many species trying to make their way around the planet. The prevailing theory of Neanderthals' demise is that our more dominant lineage effectively destroyed them. Except that they still do exist. One estimate is that 20 percent of Neanderthal DNA survives in modern humans. The victors write the history books, and so we call ourselves human, not Neanderthal.

If we step back through history, we can see that what it means to be a human has changed radically over time. When we think of the beginning of human history, we usually think back to the origin of human writing, which first developed in Mesopotamia about 5,000 years ago. We think of the worlds' religions, languages, and cultures as having been developed within this relatively narrow band of time. Even Judaism, a tradition we usually think of as ancient and eternal, only cohered into a proper religion around 2,500 years ago, when the Torah was amalgamated from beliefs and stories from previous traditions.

In fact, the history of *homo sapiens* begins around 200,000 years ago, give or take a hundred thousand years. As early as 40,000 years ago, humans created art, made stone tools, fired clay figurines, made clothes and jewelry, and developed rudimentary forms of language. Human culture reaches back in time at least eightfold beyond what we usually think of as ancient history.

The progress of science continually recalibrates the centrality of humans in the world. Prehistoric cultures believed their territory was the center of the Universe. Hawaiians' creation story involves Madame Pele, the fire goddess, who resides on Mauna Kea along with Na-maka-o-Kaha'i, the sea goddess, and Poli'ahu, the goddess of snow. (Only in Hawaii could such a creation myth exist, because only there do ocean, snow, and fire coexist).

In Europe, the old geocentric model of the Universe with earth at its center was overturned by Copernicus's heliocentric model in the sixteenth century. And eventually this model, with the sun at the center of the Universe, was itself overturned. As early as 1584, the Italian mystic Giordano Bruno proposed an infinite universe with numerous suns, around each of which move "earths." By the nineteenth century, astronomical observations had confirmed that the sun is not in fact the center of the Universe, but is one of several hundred billion stars in the Milky Way galaxy.

For the next century, astronomers believed that the Milky Way comprised all there is of the Universe. But this galactic-centric view was overturned, too, in 1920, when Edwin Hubble showed that our galaxy is just one of many in the Universe. And now multiverse theory, if true, suggests that even our seemingly unique universe may be one of many — perhaps an infinite number — of worlds.

Just as science is forever expanding our perspective about the Universe and Earth's small place in it, it is also expanding our perspective about where humans stand in the vast cosmic reality. The question of extraterrestrial life has captivated humans since far before recorded history. In ancient Greece, Anaximander speculated that there may be many other worlds besides Earth, though that line of thinking was squashed by the Earth-centric beliefs of Aristotle and Plato. Early Christians suppressed the idea of other worlds, but later theological thought occasionally admitted that the possibility of other worlds was not necessarily in conflict with Biblical teachings. The twelfth-century Islamic theologian and philosopher Fakhr al-Dīn al-Rāzī interpreted the Quranic verse "All praise belongs to God, Lord of the Worlds" to indicate other worlds or even other universes.

Recent developments in astronomy have demonstrated that the Universe is far more fecund than we'd ever guessed — some estimates set the number of potentially habitable planets at 160 billion. These are scattered across up to 3 *trillion* galaxies, containing as many as 5 sextillion stars — a 5 with 21 zeroes after it. And these galaxies are connected in even larger megastructures, superclusters hundreds of millions of light-years across. The largest actual object ever observed (apart from the Universe itself) is called the Sloan Great Wall, a filament of galaxies a billion light years across. These numbers are, quite literally, unimaginable. Our brains are not constructed to easily conceive of what 5 sextillion stars looks like.

So with billions and billions of appealing planets to choose from, the prospect of intelligent life somewhere in the Universe seems reasonable. But *how* reasonable? What are the actual odds of intelligent life existing somewhere besides Earth?

The most famous formula to estimate the odds that E.T. exists is called the Drake Equation, created by the astronomer Frank Drake in 1961. It consists of a number of parameters that are multiplied together to determine the likelihood that intelligent life is out there somewhere. Some of these parameters are relatively easy to estimate, but others are really just wild guesses, and may forever remain so. For example, astronomers are able to estimate the number of habitable planets in the Universe with increasing accuracy. But no one knows how how many of those planets will develop intelligent life. We only know of one species, on one planet, that has ever made the mysterious leap to self-awareness. We don't know if this is commonplace, or if it's extraordinarily special – or even unique.

Even so, many scientists believe that, with so many possibilities, there *should* be intelligent civilizations out there. The question is: Where are they? This is the essence of the so-called Fermi Paradox, which asks: if there is such a high likelihood of intelligent life in the Universe, why has it not yet contacted us?

The proposed explanations for the Fermi Paradox indicate much about our hopes, fears, and cares as humans. They range from the mundane (our estimates of the rate of star formation are way off) to the depressing (any civilization that achieves the ability to communicate across space will quickly extinguish itself) to the teleological (the Universe is designed to create just one form of intelligent life, and it is us). One adventurous theory, the "zoo hypothesis," proposes "zookeeper" civilizations which keep us under careful surveillance in order to protect our development, in much the way we preserve natural areas for animals to flourish shielded from human impact.

Whether extra-terrestrial intelligence exists is not just a question for space nerds and woo-woo New Agers. It has profound implications for our very understanding of what it means to be human. If there is other life out there, it means we share the Universe with an unimaginable cosmic zoo. What do those aliens think about philosophy, or science, or God? Is their awareness similar to ours, and do they perceive themselves as separate beings, or as part of the universal whole? Did their "brains" evolve to perceive time like ours, or do they exist somehow *out* of time? Does our awareness live in the same cosmic zoo as theirs? If our souls survive beyond the death of our bodies, might we find ourselves reincarnated in another galaxy?

If, on the other hand, no aliens exist – that is, if we are alone in the Universe and there is no other sentient life out there – the theological implications are profound. It means that the Goldilocks Enigma, the premise that the Universe has been perfectly designed to support the creation of life, gains even more credibility. If the Universe supports exactly one sentient race – no more, no less – that surely would be extraordinarily unlikely to have happened by chance.

The reality is that we will probably never know whether other sentient beings exist in the Universe, at least within our lifetimes. Perhaps this is for the better; any civilization that contacts us is likely to be vastly technologically superior, and the experiences of first contact, from the Spaniards in Central America to the Europeans in Africa, show us that even a small difference in knowledge or resources can have devastating consequences. Whether we are alone in the Universe or not remains a profound mystery, a sort of cosmic Zen koan – one that is wonderful to contemplate while standing on the crest of Seven Utes Mountain on a cool July morning.

AI and the future of humans

Any philosophical discussion of what it means to be human will naturally lead to questions about the future of our species. How long will the human race persist? Will we simply fade out, or will we evolve into some other life form? Or will a higher intelligence eventually overpower us, in the way that we have overpowered less sophisticated species?

Speculative science fiction has always been fascinated by the question of what happens to humans when technology allows computers to replicate or even exceed our mental functions. As technology rapidly advances, artificial intelligence is moving science fiction closer to reality. This opens up a fascinating set of philosophical questions about the nature of humans and our relationship with other self-aware life forms.

Some tools have been proposed to try to identify when a computer is "human-like." The most notable of these is the Turing Test, devised by Alan Turing, the British scientist who helped develop the field of modern computer science. It evaluates whether a computer can respond to a natural language question in a way indistinguishable from a human.

The Turing Test does not necessarily tell us much about computers and sentience, so it's become more of a philosophical tool than a meaningful goal for computer engineering. The Siri function on my iPhone can pass the Turing test, in that I can ask it a natural-language question and it can report back information that might be indistinguishable from that given by a human. But Siri is not exactly "thinking" – it's using a brute-force approach to statistically evaluate patterns of existing data created by humans.

We already know that computers can perform many "cognitive" tasks at level equal, or superior, to humans. Computers can translate languages, identify emotions, and recommend music. So the question of whether a computer can process information as well as a human has already been answered: it can.

The really interesting question is not whether computers can *compute* like humans, but whether they are *self-aware.* This is the Turing Test of our time, and it raises an interesting question: How would we ever know if a computer has, in fact, achieved self-awareness?

A computer with sufficiently nuanced algorithms could make it "appear" that it was self-aware merely by parroting what humans have already expressed. For example, if I ask a computer "What does it feel like to be conscious?" it could review passages from books that humans wrote and put together an answer. But that answer would be synthetic; it wouldn't be drawn from the computer's own qualia – its own subjective experience of awareness.

Mystical naturalism offers us a new frame with which to understand AI and consciousness. When we think about consciousness, we typically think of it in a dualistic, human-centered frame: I am a human, and I am conscious, so therefore consciousness must resemble this thing that humans experience.

Moreover, our habit is to tie consciousness to free will, and, by implication, to see it as fundamentally incompatible with deterministic behavior. A conscious being, we assume, can assert its will and tell us it is conscious. We assume that something whose behavior is obviously rigidly determined, such as a planet or a computer, cannot be conscious. A planet has no free choice; it is merely subject to the laws of nature. In its rigid deterministic behavior, how in the world could it ever express its own self-awareness?

Once we eradicate our traditional dualistic framework about the nature of ourselves and the world, the question of what might be conscious becomes completely different. Our consciousness – our deepest selves – is an expression of the laws of nature. We are the living awareness of the deterministic laws we embody. Whether or not consciousness requires free will is irrelevant; as we saw in Chapter 6, free will requires determinism, so if consciousness requires free will, it also requires determinism.

The implication is that there is no particular reason why computers, or trees, or galaxies are *not* conscious at some level. The main argument against such consciousness is simply based on analogy. Humans are conscious, and trees don't have brains like humans and don't communicate like humans, so we assume they are not conscious.

Perhaps the way a tree would communicate self-awareness is in a language we can't possibly understand. Perhaps computers will grow to develop their own language of self-awareness. Perhaps they already do.

On the other hand, it may be true, as many neuroscientists propose, that any form of consciousness must resemble human consciousness, and the structures that generate it must look like human brains. We simply may never know.

What we do know is that the idea of a "human" is something our brains have created to help us make sense of the world and survive to pass our genes along to the next generation. We are so much simpler, and so much more, than that: a glimmering light of the conscious universe.

CHAPTER 14

Mystical Naturalism and the Meaning of Life

If the whole universe has no meaning, we should never have found out that it has no meaning: just as, if there were no light in the universe and therefore no creatures with eyes, we should never know it was dark. *Dark* would be without meaning.

- C.S. Lewis

The high point of the Never Summer 100k – literally and emotionally – is the 11,587-foot peak of North Diamond Mountain. The final climb to the top is so steep I have to take two or three steps at a time, hands on knees, and then pause for breath. Ahead of me and behind me, racers repeat the same ritual as they slowly ascend the slope.

I step a few yards off the trail to pee and I'm suddenly nowhere, alone in a dense grove of trees. There's a whole world in here, seemingly unaware of the highway of humans grinding up the mountain a few feet away. I take a moment to just be here, to think about this place and its long, slow march of growth, death, and rebirth on a timeline very different from mine. It's good to step away from the ongoing drone of the 24-hour news cycle and into the news cycle of nature, so far removed from the familiar everyday stream of Trump tweets and coronavirus cases.

I gather my strength and make a final push to the summit, where a bluegrass band greets runners as they pass, playing gamely into the whipping wind. A strip of ethereal clouds covers the base of the mountains below, but the sky above is clear. I have an unobstructed 360-degree view, mountain range beyond mountain range.

The wholeness and immediacy of being in nature helps to coalesce all the strands of this journey I've been on for so long. Nature has a way of putting me in my place; of simultaneously making me feel tiny and an essential part of an unfathomable master plan. Here I find the beating heart of mystical naturalism: experiencing myself as a part of the miracle of Creation.

I've traversed high peaks and vertiginous ridgelines, scaled mountainsides, and earned breathtaking vistas. I've explored, however briefly, the workings of the Cosmos and the mysteries of the human brain; examined the evidence for God; investigated arguments for how the world came to be; and questioned my very identity as a human being. Truly understanding all of these mysteries would take innumerable lifetimes; I feel like I've just scratched the surface on my own whirlwind journey.

Across all those miles, all those excursions into quantum physics and neurobiology and consciousness research, I've come away with a profound realization, a simple but life-changing insight.

That insight is this: it is duality that obscures the truth. It is duality that gives our ego unearned power, that separates us from each other, that seeds so many supposed philosophical contradictions. When we identify and extract this villain we gain a whole new understanding of ourselves, our role in the world, and even the nature of God.

Nonduality shows us that we are not separate from the natural world; we are simply an aspect of it. It shows us that the natural world and the supernatural world are the *same* world, hidden in plain sight right in front of us, no hidden codebooks or magical thinking required.

As I cross the summit and start my long descent down the other side, the dull ache of climbing melts away. My thoughts dissipate into the thin mountain air as I focus on each footfall. It's as close as I've gotten in a long while to ecstasy.

This is what the psychologist Mihaly Csikszentmihalyi called "flow," the mental state of being completely involved in an activity for its own sake, as your ego falls away. "The purpose of the flow is to keep on flowing, not looking for a peak or utopia but staying in the flow. It is not a moving up but a continuous flowing; you move up to keep the flow going." Happiness is not a means in and of itself but an "unintended side-effect of one's personal dedication to a course greater than oneself."

The existence of flow reflects the brute fact that the world must be just how it is, and to find happiness we must fully immerse ourselves in it. But this is different than a nihilistic submission to whatever life brings us. In fact, happiness requires a balance between accepting what is and envisioning what could be.

A scientific perspective on this balance comes from the positive psychology movement, which grew from the work of Martin Seligman and other psychologists in the late 1990's. It focuses on the power of hope, optimism, and positive thinking to make our lives better.

The self-help-industrial complex has seized upon the findings of the positive psychology movement, sometimes accurately, sometimes not. The simple idea that positive thinking has beneficial outcomes has become jumbled up with pseudoscience, parapsychology, and New Age platitudes. One of the best-known formulations is the "Law of Attraction," coined by Rhonda Byrne in her 2006 mega-bestseller *The Secret*. The idea is reflected in the Biblical quotation "And all things, whatsoever ye shall ask in prayer, believing, ye shall receive."

The Law of Attraction has been criticized by writers like Barbara Ehrenreich, who complain that it does not address structural issues like poverty that inherently create inequity. If achieving what I want is simply a matter of visualizing it, are people who are disadvantaged or oppressed simply not visualizing hard enough? The Law of Attraction also seems to invoke a certain type of magic that somehow evades everything we know about how the physical world works. Merely by believing, it seems to say, we can manifest a $10 million check or the perfect husband or the life of our dreams.

But in spite of the pseudoscientific sheen of the Law of Attraction, many research studies show that visualization can be powerful in driving future behavior. The Law of Attraction is real, but it does not work by magic. It works in concert with the ordered processes of our brains, operating by the real rules of the real world.

In fact, it was Mahatma Gandhi, not exactly a privileged self-help guru, who said, "Every moment of your life is infinitely creative and the Universe is endlessly bountiful. Just put forth a clear enough request, and everything your heart desires must come to you."

The trick is to align what your heart desires with the facts of the real world. Visualizing possibilities is not like a buying a lottery ticket and hoping for an unearned chance at an arbitrary payout. It comes from a deep understanding of your essential place within the real world, a love of your tiny part of the unity we call reality.

This is the best definition I can think of for enlightenment: not an abstract higher state that only a precious few of us attain; but simply the embracing of the world as it truly is, right now. The ability to achieve this sort of enlightenment is not an indicator of moral righteousness or spiritual superiority; it is a skill, built through practice. All it really requires of us is trust in ourselves and in our ability to understand the world around us.

The meaning of life is to embrace the reality of the world and your tiny part of it. It is to live your story. It is to be here, now. It is to change what you can, to accept what you cannot change, and to have the wisdom to know the difference. It is mindfulness and presence. It is flow. It is to love yourself and your unbreakable connection with the rest of the Universe.

The fourteenth-century Persian poet Hafiz wrote, "The place where you are right now God circled on a map for you. Wherever your eyes and arms and heart can move against the earth and sky, the Beloved has bowed there. Our Beloved has bowed there knowing you were coming." To translate this lovely bit of poetry into more prosaic terms, the Universe simply is, across all times and out of time. It is order, it is magic, it is nature, it is God. And it is us.

References

Chapter 1: An Introduction to Scientific Seeking

just .00036 degrees above absolute zero:
   https://www.iflscience.com/physics/coldest-temperature-
   universe-created-american-laboratory/
summitted Everest ten times:
   http://articles.latimes.com/2010/nov/28/nation/la-na-
   hometown-boulder-20101128
an individual's spiritual evolution:
   http://www.rigpawiki.org/index.php?title=Sogyal_Rinpoche
active and practicing Catholic:
   https://www.csmonitor.com/Technology/2013/0219/Copern
   icus-and-the-Church-What-the-history-books-don-t-say
Christian theology, he argues:
   https://www.nytimes.com/2007/11/24/opinion/24davies
better than your superturtle:
   https://www.edge.org/conversation/paul_davies-taking-
   science-on-faith
ephemeral and changing world:
   http://faculty.evansville.edu/tb2/trip/cratylus.htm
when it comes to gains: Amos Tversky and Daniel Kahneman, "The
   Framing of Decisions and the Psychology of Choice," Science
   211(4481) (1981): 453–458.

Chapter 2: The Wise Universe

the One and the All: http://www.bible-researcher.com/logos.html
an external cause: Ethics, part III, proposition 13, sholium; in
   Frankfurt, p. 39
prayers offered in this place: 2 Chronicles 7:14-15

rebalance the energy body by removing blockages:
> https://www.energymuse.com/blog/chakra-stones-chart-chakra-awareness/

process of logical operations:
> https://schoolbag.info/mathematics/real/12.html

using made-up symbols in a consistent way:
> https://core.ac.uk/download/pdf/82047627.pdf, p. 32

inopportune lawlessness: The Story of Philosophy by Will Durant (1938). New Revised Edition, Section: Kant and German Idealism, Sub-Section: Transcendental Analytic, Quote Page 295 and 296, Published by Garden City Publishing Company, Garden City, New York. See The Quote Investigator: https://quoteinvestigator.com/2015/05/19/wisdom/#note-11193-2

mingling is own nature with it:
> https://www.goodreads.com/quotes/110897-the-idols-of-tribe-have-their-foundation-in-human-nature

course of self-improvement: https://fs.blog/2016/05/francis-bacon-four-idols-mind/

we possess but fragments of it: Spinoza, B., Shirley, S., & Feldman, S. (2001). *Theological-political treatise,* chapter 12. Indianapolis: Hackett Pub. Co.

affecting health guidelines and treatment practices:
> https://www.sciencemag.org/news/2018/08/researcher-center-epic-fraud-remains-enigma-those-who-exposed-him

careful publication guidelines have been followed:
> https://www.newyorker.com/magazine/2010/12/13/the-truth-wears-off

your mind, character, thoughts, and feelings: Collins English Dictionary – Reference Edition. (2016). Collins Dictionaries.

curable through exorcism:
> https://www.ncbi.nlm.nih.gov/pubmed/25208453

Chapter 3: Quantum Theory and the Power of Naturalism

Einstein called these little units of light quanta:
https://www.pitt.edu/~jdnorton/teaching/HPS_0410/chapte
rs/quantum_theory_origins/

the behavior of particles is truly random:
https://blogs.scientificamerican.com/observations/photons-
quasars-and-the-possibility-of-free-will/;
http://news.mit.edu/2018/light-ancient-quasars-helps-
confirm-quantum-entanglement-0820

a super-deterministic cosmic conspiracy:
https://arstechnica.com/science/2014/02/is-entanglement-is-
real-or-is-there-a-super-deterministic-cosmic-conspiracy/

fundamentally change the outcome of the game:
http://www.johnchilds.net/cueball.htm

most likely not be turned into wine: http://www.the-
brights.net/vision/essays/dennett_nyt_article.html

Chapter 4: Mystical Naturalism

you are that: http://upanishads.org.in/stories/shvetaketu

attributed to a divine agency:
https://en.oxforddictionaries.com/definition/miracle

the fact that it is comprehensible is a miracle: 'Physics and
Reality', Franklin Institute Journal (Mar 1936). Collected in Out
of My Later Years (1950), 60.

continue still:
https://en.wikisource.org/wiki/Page:Humanimmortality00jam
e.djvu/36

booming, buzzing confusion of pure experience:
https://www.britannica.com/topic/neutral-monism

Chapter 5: The Mystery of Consciousness

named after him:
https://onlinelibrary.wiley.com/doi/10.1002/9781119132363.ch1

to be in pain: https://www.nybooks.com/daily/2018/03/13/the-consciousness-deniers/

mathematically sound principles:
https://www.extremetech.com/extreme/181284-human-consciousness-is-simply-a-state-of-matter-like-a-solid-or-liquid-but-quantum

people in pictures:
http://users.wfu.edu/masicaej/Baumeisteretal2011AnnRevPsych.pdf

size and complexity:
https://www.sciencemag.org/news/2016/01/human-brain-big-internet

best *possible universe:* https://www.lrb.co.uk/v20/n03/derek-parfit/why-anything-why-this

"wants" to be conscious: https://www.prospectmagazine.co.uk/arts-and-books/thomas-nagel-mind-and-cosmos-review-leiter-nation

biased towards the marvelous: https://www.thenation.com/article/do-you-only-have-brain-thomas-nagel/#?page=full

well, fine, and good:
https://whyevolutionistrue.wordpress.com/2013/05/14/irresponsible-journalism-the-chronicle-of-higher-education-goes-to-bat-for-woo-driven-evolution/

once-great thinker: https://www.prospectmagazine.co.uk/arts-and-books/thomas-nagel-mind-and-cosmos-review-leiter-nation

replaced every year:
https://www.npr.org/templates/story/story.php?storyId=11893583

a complete replacement every month: http://book.bionumbers.org/how-quickly-do-different-cells-in-the-body-replace-themselves/

why the shovel was chosen; Michael Gazzaniga and Joseph DeLoux, 1978

reducing anxiety and mental stress:
> https://www.health.harvard.edu/blog/mindfulness-
> meditation-may-ease-anxiety-mental-stress-201401086967

can't visualize seven-dimensional space:
> http://www.pragmatismtoday.eu/summer2015/07%20Campb
> ell.pdf

Chapter 6: Determinism and Free Will

why future things will be: Cicero, On divination 1.125-6, trans. Long
    and Sedley 1987, 55L

be present before its eyes: Laplace, Pierre Simon, A Philosophical Essay
    on Probabilities, translated into English from the original
    French 6th ed. by Truscott, F.W. and Emory, F.L., Dover
    Publications (New York, 1951) p.4

most of the work has already been done:
> https://www.wired.com/2008/04/mind-decision/

choice is simply an illusion: This Idea Must Die, p. 153

Chapter 7: Why Does the World Exist?

billions of years ago: https://www.smithsonianmag.com/smithsonian-
    institution/how-scientists-confirmed-big-bang-theory-owe-it-
    all-to-a-pigeon-trap-180949741/

between hydrogen atoms: https://www.big-bang-theory.com/

a place to set his pipe:
> http://groups.creighton.edu/sjdialogue/documents/articles/ul
> timate_reality.html

Chapter 8. What is God?

must have had a cosmic designer:
> https://www.nytimes.com/2006/10/22/books/review/Holt.t.
> html

:sometimes also as saints: "Religion and Science," New York Times Magazine, November 9, 1930

how DARE you:

> https://www.theguardian.com/science/2015/jun/09/is-richard-dawkins-destroying-his-reputation

You alone are God:

> http://nisargadatta.net/Ramana_Maharshi_David_Godman_1.html

the God of my childhood:

> http://www.stsci.edu/~rdouglas/publications/suff/suff.html

Human beings did it: https://crownheights.info/general/2897/elie-wiesel-on-his-beliefs/

Chapter 9: Religion and its Discontents

I cannot see anything "chosen" about them Gutkind Letter (3 January 1954), "Childish superstition: Einstein's letter makes view of religion relatively clear". The Guardian. 13 May 2008.

liberal American Jews: https://www.haaretz.com/jewish/.premium-israel-to-reform-jews-youre-not-jewish-but-your-money-is-1.5376021

part of reality too: https://biologos.org/common-questions/how-is-biologos-different-from-evolutionism-intelligent-design-and-creationism/

violent enmities in the world today:

> https://www.goodreads.com/author/quotes/1194.Richard_D awkins?page=2

relationship is nuanced: Green and Elliot (2010)

caused more healthy behaviors:

> https://www.jstor.org/stable/2648114?seq=1#page_scan_tab _contents

to the American economy:

> https://www.forbes.com/sites/jeffreydorfman/2013/12/22/r eligion-is-good-for-all-of-us-even-those-who-dont-follow-one/#147eb24164d7

better health, and stronger marriages:
https://www.amazon.com/Americas-Blessings-Religion-
Benefits-
Including/dp/159947445X/ref=sr_1_8?ie=UTF8&qid=13876
44284&sr=8-8&keywords=rodney+stark

strong social ties there: https://time.com/collection/guide-to-
happiness/4856978/spirituality-religion-happiness/#

engaged in other social groups: https://www.pewresearch.org/fact-
tank/2019/01/31/are-religious-people-happier-healthier-our-
new-global-study-explores-this-question/

better social support:
https://www.psychologytoday.com/us/blog/supersurvivors/2
01809/is-religion-good-or-bad-us

less happy than atheists and agnostics:
https://www.hbs.edu/faculty/Pages/item.aspx?num=40219

stress, shame, depression, and anxiety:
https://www.abc.net.au/news/2017-10-19/anti-gay-religious-
prejudice-study-mental-health/9061606

a prescription for living and acting rightly:
https://tricycle.org/magazine/who-was-buddha-2/

a practicing Buddhist: https://www.linkedin.com/pulse/yuval-noah-
harari-closet-his-religion-angus-douglas-/

in a controversial:
https://www.theguardian.com/world/2005/jul/27/research.hi
ghereducation

as found in ancient Buddhist texts:
https://www.dalailama.com/messages/buddhism/science-at-
the-crossroads

dipping their hands into water: https://tricycle.org/magazine/the-five-
precepts/

beyond religion altogether: The Dalai Lama's Facebook statement for
September 10, 2012

Chapter 10: Discovering Morality

the Against Malaria Foundation, at $3,337:
> https://www.businessinsider.com/the-worlds-best-charity-can-save-a-life-for-333706-and-thats-a-steal-2015-7

innate sexual preferences: https://www.thetrevorproject.org/get-involved/trevor-advocacy/50-bills-50-states/conversion-therapy-research/#sm.0001vw98h6ufjd7wwdv1kb5lx3or6

Chapter 11: Life After Death and Psi

occur in hospitals:
> https://www.cdc.gov/nchs/products/databriefs/db118.htm

for his experiences:
> https://www.esquire.com/entertainment/interviews/a23248/the-prophet/

insists on a supernatural one:
> https://michaelshermer.com/2013/04/proof-of-hallucination/#more-4183·

only a few experimenters: Radin (1997), p. 144, in
> http://skepdic.com/refuge/radin9.html;
> http://skepdic.com/pear.html

myself and no one else: The New Science and Spirituality Reader, p. 13

Chapter 12. The Mystery of Time

infinitely far into the past:
> www.preposterousuniverse.com/writings/dtung

a stubbornly persistent illusion: Einstein, A., 1955, 'Letter to the family of Michele Besso, after learning of his death', quoted in Dyson, F. J., 1979, 'Disturbing the universe', Basic Books.

a seamless, continuous movie: This Idea Must Die, p. 307

unless you're planning to use it in the future:
> https://www.nytimes.com/2007/04/03/science/03time.html

https://www.nytimes.com/2017/05/19/opinion/sunday/why-the-future-is-always-on-your-mind.html

fear, hope, anxiety, worry: https://www.thoughtco.com/about-time-449562

Chapter 13: What is a Human?

born as 46XX:
  https://www.who.int/genomics/gender/en/index1.html
more like labia: http://www.isna.org/faq/what_is_intersex
rudimentary forms of language:
  https://australianmuseum.net.au/learn/science/human-
  evolution/homo-sapiens-modern-humans/
aspire to the greater good: https://americanhumanist.org/what-is-
  humanism/definition-of-humanism/
Harari, *Sapiens*, p 230
Protestant settlers believed: Science of Happiness magazine, p. 85

Chapter 14: Mystical Naturalism and the Meaning of Life

trajectory of the ball:
  http://users.wfu.edu/masicaej/Baumeisteretal2011AnnRevPsy
  ch.pdf
more frequent health checkups: New York Times, July 15, 2018

Bibliography

Abrams, N.E. (2015). A God That Could Be Real. Beacon Press.

Carroll, S. (2017). The Big Picture: On the Origins of Life, Meaning, and the Universe Itself. Dutton.

D'Souza, D. (2009). Life After Death: The Evidence. Salem Books.

Davies, P. (1993). The Mind of God: The Scientific Basis for a Rational World. Simon and Schuster.

Dawkins, R. (2008). The God Delusion. Mariner Books.

Feyerabend, P. (1978). Science in a Free Society. Verso.

Feyerabend, P. (2010). Against Method. Verso.

Frankfurt, H. (2006. On Truth. Knopf.

Harari, Y.N. (2016). Sapiens: A Brief History of Humankind. Harper Perennial.

Harris. S. (2012). Free Will. Free Press.

Hawking, S. (1988). A Brief History of Time: From the Big Bang to Black Holes. Bantam.

Holt, J. (2013). Why Does The World Exist?: An Existential Detective Story. Liveright.

Horgan, J. (2004). Rational Mysticism: Spirituality Meets Science in the Search for Enlightenment.

Kahneman, D. (2013). Thinking, Fast and Slow. Farrar, Straus, and Giroux.

Keller, The Reason for God: Belief in an Age of Skepticism. Penguin Books.

Lanza, Beyond Biocentrism: Rethinking Time, Space, Consciousness, and the Illusion of Death (2017). BenBella Books.

Parfit, D. (1986). Reasons and Persons. Oxford University Press.

Peterson, M, Hasker, W., Reichenbach, B., Basinger, D. (2012). Reason and Religious Belief: An Introduction to the Philosophy of Religion. Oxford University Press.

Pirsig, R. M. (1974). Zen and the Art of Motorcycle Maintenance: An Inquiry Into Values. William Morrow Paperbacks.

Pirsig, R. M. (1992). Lila: An Inquiry Into Morals. Bantam.

Pollan, M. (2019). How To Change Your Mind: What the New Science of Psychedelics Teaches Us About Consciousness, Dying, Addiction, Depression, and Transcendence. Penguin Books.

Selbie, J. (2017). The Physics of God: Unifying Quantum Physics, Consciousness, M-Theory, Heaven, Neuroscience and Transcendence. Weiser.

Shermer, M. (2016). The Moral Arc: How Science Makes Us Better People. St. Martin's Griffin.

Singer, P. (2009). The Life You Can Save: How to Do Your Part to End World Poverty. www.thelifeyoucansave.org.

Suzuki, Shunryu (1970). Zen Mind, Beginner's Mind: Informal Talks on Zen Meditation and Practice. Shambhala.

Wade, N. (2010). The Faith Instinct: How Religion Evolved and Why it Endures. Penguin Books.